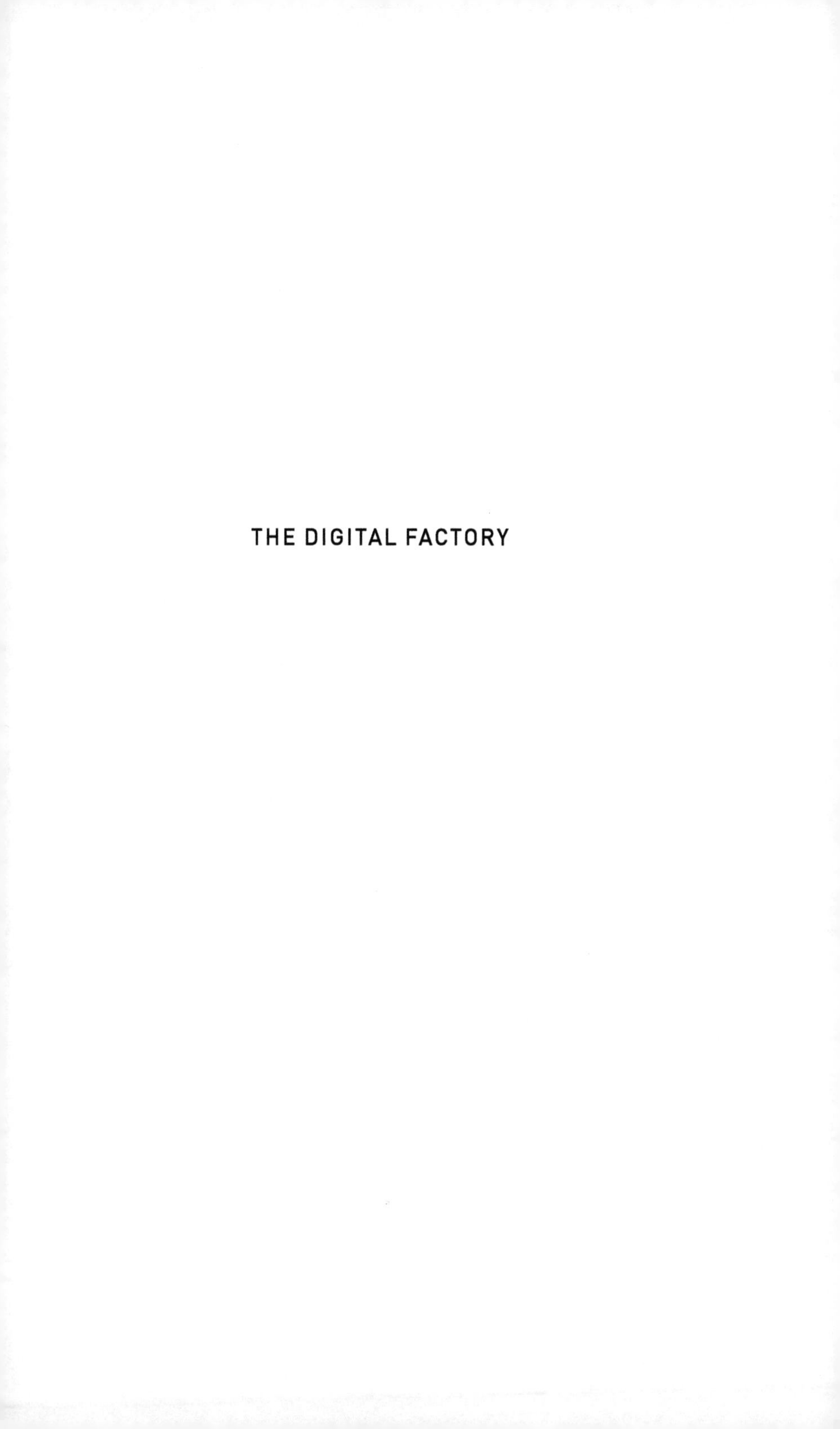

THE DIGITAL FACTORY

THE DIGITAL FACTORY

THE HUMAN LABOR OF AUTOMATION

MORITZ ALTENRIED

The University of Chicago Press
CHICAGO AND LONDON

The University of Chicago Press, Chicago 60637
The University of Chicago Press, Ltd., London

Published 2022
Printed in the United States of America

31 30 29 28 27 26 25 24 23 22 1 2 3 4 5

ISBN-13: 978-0-226-81549-7 (cloth)
ISBN-13: 978-0-226-81548-0 (paper)
ISBN-13: 978-0-226-81550-3 (e-book)
DOI: https://doi.org/10.7208/chicago/9780226815503.001.0001

Library of Congress Cataloging-in-Publication Data

Names: Altenried, Moritz, author.
Title: The digital factory : the human labor of automation / Moritz Altenried.
Description: Chicago : University of Chicago Press, 2022. | Includes bibliographical references and index.
Identifiers: LCCN 2021021841 | ISBN 9780226815497 (cloth) | ISBN 9780226815480 (paperback) | ISBN 9780226815503 (ebook)
Subjects: LCSH: Digital labor. | Internet industry—Employees. | Employees—Effect of technological innovations on. | Labor supply—Effect of technological innovations on. | Industrial management—Technological innovations. | Technological innovations—Social aspects. | Technological innovations—Economic aspects.
Classification: LCC HD9696.8.A2 A47 2022 | DDC 338.4/7004678—dc23
LC record available at https://lccn.loc.gov/2021021841

CONTENTS

1

WORKERS LEAVING THE FACTORY

Introduction

The Googleplex is Google's Silicon Valley headquarters. Located in Mountain View, California, it consists of an assortment of glass and steel buildings with colorful features in the company's colors sprinkled in. The complex sprawls out over a large area of land and continues to grow as Google steadily adds new sites and buildings to its headquarters. Between the buildings, most of which are of only medium height, are many green areas, parking spaces, and recreational facilities. Amenities of the complex include free restaurants and cafeterias, four gyms, swimming pools, beach volleyball courts, cinemas, and lecture halls.

An artist's video installation from 2011 called *Workers Leaving the Googleplex* engages with this complex and its architecture of labor.[1] The video work by Andrew Norman Wilson shows various buildings, parks, and cafés of the complex used by Google's employees. On the left side of the screen, individuals and small groups can be seen entering or leaving the buildings from time to time. Some use one of the free bikes Google supplies to its employees, while others eat in one of Google's free gourmet cafeterias after a day of working and before boarding the luxurious shuttle bus back to San Francisco after their meal. Google operates over one hundred such buses, which are equipped with wireless internet access and other amenities to shuttle its employees from around the Bay Area to the complex.

Like the headquarters of other prestigious information technology companies, the Googleplex's design is inspired more by a university campus than traditional office or factory buildings. A promotional video describes the Googleplex as "very academic," but also "a big playground" with an "eccentric" atmosphere.[2] The architecture of the buildings on the Mountain View campus corresponds to the firm's understanding of work. The terms with which Google describes the Googleplex as a workplace include words like *freedom*, *creativity*, *flat hierarchies*, *playful*, *communicative*, and *innovative*. The buildings are designed to bring the "Googlers" in touch with one another—private offices are a rare occurrence, and employees are encouraged to pursue their own projects during working hours. At the Googleplex's main buildings, no masses leave at the same time, and no shift changes can be found—only individuals or groups dropping in and out, seemingly at their leisure. Is this the quintessence of labor in digital capitalism? According to Wilson's video—and this book—not quite.

Workers Leaving the Googleplex consists of a split screen. The left side displays the images just described, while the other side shows a very different set of workers at Google. Wilson, a contractor in Google's video department at the time, accidentally discovered another type of Googler working in a building next door, displayed on the right side of his video project's split screen. What initially piqued Wilson's curiosity was that these workers left their building in one large group. Unlike the Googlers in the prestigious main buildings, the workers in this inconspicuous adjacent building do in fact work in shifts and can be identified by their yellow badges. Google's employees are split into various groups marked by visible badges everyone is obliged to wear—for example, white badges are for full-time Googlers, Wilson wore the red badge designating outside contractors, and green badges are for interns.

These "yellow badges" worked for Google's controversial project of digitizing every book in existence. In 2010, Google estimated that 130 million unique books existed in the world and announced plans to digitize all of them by the end of the decade. Although technology has improved significantly in recent years, the process is still not fully automated. This, in turn, created the need for the book-scanning labor of the yellow badges, referred to as "ScanOps." While other digitizing projects have reduced labor costs by shipping containers full of books to be scanned in India and China, Google employed the services of subcontracted workers at its Mountain View facilities.

They work in shifts. The ones Wilson filmed began at 4:00 a.m. and left the Google Books building at precisely 2:15 p.m. Their work consists of turning pages and pressing the scan button on a machine. A former worker describes his experience: "I had a set amount of instructions to follow and a certain quota I was to meet every day. The only thing that changed were the books we had to scan and the quality in which we got those books."[3] Google has developed its own scanning technology and a patented machine that instructs the worker to turn the pages timed to a rhythm-regulated soundtrack. Wilson soon learned that this group of workers lacked the aforementioned employee privileges such as free meals, access to gyms, bikes, shuttle services, free presentations, and cultural programs. They are not even allowed to move freely around the Google campus, and Google does not like them talking with other employees—as Wilson found out while making his film. Google's security stopped him from filming and interviewing the yellow badges, and he was subsequently fired for his investigation.

Wilson's split-screen installation is a recognizable reference to two earlier works on the architecture of labor, the first being Louis Lumière's *Workers Leaving the Lumière Factory*, often described as the first real motion picture ever made. It consists of a 46-second shot of the Lumière factory gates in Lyon, France, and depicts the mostly female workforce leaving the Société Anonyme des Plaques et Papiers Photographiques Antoine Lumière, a successful photograph manufacturer. Originally shot in various versions in spring 1885, the scene of workers leaving the factory gates has inspired numerous remakes and new versions, most famously Harun Farocki's 1995 *Workers Leaving the Factory*, in which shots of the Lumière factory are spliced into an assemblage of workers leaving factories across various locations and historical periods.

To me, this seemingly untimely reference to the factory is intriguing. The factory has long been a symbol of economic and social progress as well as a central point of departure for a critical analysis of capitalist societies. Crucial for the making of the modern world, giant factories were, for quite some time, viewed "as templates for the future, setting the terms of technological, political and cultural discussion," according to historian Joshua Freeman in his book *Behemoth*, which delves into the history of the factory.[4] From the English factories featured in Marx's *Capital* to the Ford factories that became the namesake of an entire period of capitalism—for over a century, the factory was also at the heart of many critical economic and social theories as well as political practice.

If the factory is central to theories of Fordism, then theories of post-Fordism are mostly postfactory theories. Sidelined in the depictions of the post-Fordist variant of capitalism, the factory's most important role is often that of a counterexample against which the transformation of labor and capitalism is analyzed. Hence, it seems that the factory has rapidly lost the central role it has held in most understandings of capitalism for over a century. In contrast to this, a central approach guiding this book is to think about forms of continuity of the factory in order to understand contemporary digital capitalism.

At this writing, it appears Google's enthusiasm for scanning every book available has declined and that the project has decreased in size and importance. The model by which it is run, however, has proliferated. Google's workforce includes a contingent of over one hundred thousand workers designated as TVCs (temps, vendors, and contractors), such as the book scanners. These subcontracted workers transcribe, for example, conversations to train Google's digital assistant, drive cars capturing photos for Google street view, and monitor videos uploaded to YouTube for dangerous content. Some of them work directly next to the high-paid Googlers on the Mountain View campus, others work in call centers around the world, and yet others in their private homes. If possible, Google avoids talking about them and hides them behind brick walls and digital interfaces. Often, they carry out work that most people assume is being performed by algorithms. Not only at Google but elsewhere, these workers are a crucial yet often overlooked part of contemporary digital capitalism. Hence, the following chapters concentrate on digital factory sites such as the inconspicuous building adjacent to the Googleplex. Although these digital factories differ significantly, they are all sites in which we find labor regimes that have little in common with the creative, communicative, or glamorous image of work at Google's main building.

INTO THE DIGITAL FACTORY

This is a book about the transformation of labor in digital capitalism. Centrally, it addresses the impact of digital technology, particularly sites where it brings forth labor relations characterized by features one might assume only exist in traditional factories. The focus on such sites of labor opens a particular perspective on the transformation of labor and capitalism in the digital age. Many important critical theories of the transformation of labor have highlighted its creative, communicative, immaterial, or artistic features. Furthermore, contemporary discussions often contain arguments that digital technology and automation are doing away with forms of menial and routinized labor. Without denying the importance of, for example, creative labor for the contemporary or ongoing processes of far-reaching automation, I argue that this process is neither uniform nor linear and turn to sites where the impact of digital technology has enforced different developments.

Hence, the interest in workers like the yellow badges described in the opening scenario. This investigation covers sites that may not always look like factories, but where the logics and workings of past factories are very much present, often accelerated by the increasing pervasiveness of digital technology. Whether Google's scanning workers in California, crowdworkers or warehouse workers in Germany or Australia, gaming workers or content moderators in China or the Philippines, Deliveroo drivers or search engine raters in the UK or Hong Kong, video game testers or Uber drivers in Berlin or Nairobi: these are the workers of today's digital factory. Repetitive yet stressful, often boring yet emotionally demanding, requiring little formal qualification yet oftentimes a large degree of skill and knowledge, and inserted into algorithmic architectures not yet automatable (at least for now)—these segments of labor are a crucial part of the political economy of the present. This book investigates sectors of labor in which digital technology ensures and enforces labor regimes sometimes curiously resembling those of Taylorist factories in the early twentieth century, even if they look completely different. It examines sites where the development of digital technology requires human labor in forms that are highly fragmented, decomposed, and controlled. These areas of work are often hidden behind the magic of algorithms, thought to be automated but in fact still highly dependent on human labor—they are, effectively, *digital factories*. This analysis of such digital factories and the workers employed there, their technical and political composition, new forms of labor organization, their mobility and migration practices, and the emergence of new geographies of production and conflict is aimed to contribute to a better theoretical and empirical understanding of the contemporary moment, shaped by the encounter of globalized capitalism and digital technology.

A central argument of this book is, then, that digital capitalism is not characterized by the end of the factory, but by its explosion, multiplication, spatial reconfiguration, and technological mutation into the digital factory. Essentially, the factory is a system of organizing and governing the production process and living labor.[5] In this sense, the factory is understood as both a real site of labor as well as—more abstractly—an apparatus and logic for the ordering of labor, machinery, and infrastructure across space and time. The reconfiguration of this process by digital technology is the focus of this book. This entails less emphasis on the ongoing importance of industrial factories or the digitization and automation of manufacturing in these factories (something discussed using, for example, the buzzwords "fourth Industrial Revolution" or "Industry 4.0") than on the digital factory: a search for how digital technology transforms how labor is organized, composed, and distributed spatially. It sets out to describe how digital technology allows the logic of factories to find new spatial forms, such as the platform.

In its analysis of the transformation of labor, the book develops three central vectors that return throughout the chapters and different sites of the investigation. As a first vector, the term *digital Taylorism* is developed both empirically and theoretically throughout the book. Digital technology has manifold implications for the transformation of labor; digital Taylorism is only one of many ways manufacturing management has transferred to the digital world. While many journalistic and academic works investigating the implications of digital technology for the world of work are concerned with possibilities of automation or the increasing immateriality of labor, the concept of Taylorism has also seen a small resurgence.[6] Today, the term is mostly used polemically, rarely systematically, to describe how digital technology allows for new modes of workplace surveillance, control, and deskilling. *The Economist*'s Schumpeter column even muses that "digital Taylorism looks set to be a more powerful force than its analogue predecessor."[7]

I use the term to describe how a variety of forms and combinations of soft- and hardware as a whole allow for new modes of standardization, decomposition, quantification, and surveillance of labor—often through forms of (semi-)automated management, cooperation, and control. Even if digital technology allows for the rise of classical elements of Taylorism such as rationalization, standardization, decomposition, and deskilling, as well as the precise surveillance and measurement of the labor process, this is not a simple return of Taylorism; rather, the phenomenon has emerged in novel ways. Thus, by invoking Taylor, I do not argue for a simple rebirth of Taylorism but rather seek to emphasize how digital technology allows for the rise of classical elements of Taylorism in often unexpected ways. These forms of algorithmic management and control of the labor process allow for new forms of the subsumption of labor under capital outside the traditional factory. In many ways, digital technology is able to take on the spatial and disciplinary functions of the traditional factory and develop new forms of coordination and control that can reach out onto streets or into private homes.

In analyzing and conceptualizing today's digital capitalism, this book turns to a new perspective: It focuses less on areas where small groups of workers supervise the operation of machines than on areas of work characterized by algorithms organizing the labor of large numbers of human workers. It addresses not so much the creative and communicative elements of computerized labor as it does the fragmented, controlled, and repetitive segments (and, in turn, their forms of creativity and communication). Less concerned with the future impact of artificial intelligence (AI), it instead observes the workers training this AI today. An approach proceeding from these forms of digital Taylorism opens up a perspective less geared toward projections of how digitally driven automation will replace living labor, instead shedding light on the complex and manifold ways living labor is reconfigured, newly divided, multiplied, and displaced in the contemporary world.

Concerning the composition of labor, these complex dynamics can be described within the analytical framework of the *multiplication of labor*, a second vector of the analysis laid out in this book. The digital factory can articulate different workers without homogenizing them in spatial or subjective terms. Herein lies a crucial difference to traditional Taylorism, namely, the digital factory produces no digital mass worker like the industrial mass worker. Digital technology, or, more precisely the standardization of tasks, the means of algorithmic management, and surveillance to organize the labor process, as well as the automated measuring of results and feedback allow for the inclusion of a multiplicity of often deeply heterogeneous workers in multifarious ways. It is then precisely the standardization of work (as conceptualized by the term *digital Taylorism*) that allows for the multiplication of living labor in many ways.

I use the term *multiplication of labor* building on Sandro Mezzadra and Brett Neilson's important work.[8] They use the term to supplement the familiar term *division of labor* to hint at the heterogeneity of living labor in a time characterized by the increasing coalescing of labor and life, the increasing flexibilization of labor, as well as shifting and overlapping geographies in the ongoing processes of globalization. Indeed, the concept also is extremely effective in shedding light on the transformation of labor driven by digital technology as I hope to show throughout the book. First, the concept of the multiplication of labor alludes to the fact that the digital factory allows the tightly controlled and standardized cooperation of a large number of workers who may come from different backgrounds, experiences, and locations. Whether in a distribution center or the gig economy, to take two examples from the following chapters, digital technology and automatically managed and standardized work procedures allow for the quick inclusion as well as substitutability of workers and hence contribute to the flexibilization and heterogenization of labor.

Second, throughout the book, we can observe the literal multiplication of labor in the sense that a large number of people need to work more than one job. Often, this includes a further blurring of times of labor and free time. Hereby, the flexibilization of labor and the trend toward unstable and multiple labor arrangements instead of the Fordist ideal of one stable and lifelong job can be observed at many places (although it is important to add that this ideal was achieved only for a limited time by a specific segment of the working class, limited by vectors of gender, racism, and geography, among other factors).

Third, digital technology is implicated in the reconfiguration of the mobility of labor and goods, be it through transforming logistical systems or new forms of labor migration such as the advent of virtual migration. In this sense, the multiplication of labor encompasses a specific heterogenization of labor geographies and labor mobility, a reconfiguration of the gendered division of labor, and the proliferation of flexible contractual forms such as short-term, subcontracted, freelance, and other forms of irregular employment.

It should be clear by now that space is a crucial dimension to these developments. Understanding the digital factory explicitly as a spatial concept, the reconfiguration of space through digital infrastructures serves as a third vector underlying the analysis. Digital infrastructures are profoundly reshaping the production of space in almost all areas of life, as well as the geography of labor, from the smallest detail to the geopolitical dimension. Keller Easterling's analytical reformulation of understanding infrastructure as "infrastructure space" is important not only to analyze how infrastructure and digital technology are implicated in the production of space, but even more so for how they reorganize the spatiality of labor: global logistical systems that reorganize the global division of labor, software that minutely organizes workers' movements through an Amazon warehouse, or crowdworking platforms bringing digital labor into private homes across the globe are examples for how digital technology changes the spatial architecture of labor.[9]

If digital technology is able to move the factory (as a labor regime) beyond the factory (as a concrete building), the digital factory can take different spatial forms. Those include, for example, the platform. Like the traditional factory, today's digital platforms (e.g., Uber, Amazon Mechanical Turk) organize the labor process and social cooperation across space and time. The infrastructure facilitating such processes is of paramount importance, both for the concrete functioning of the digital factory as well as for the reconfiguration of economic space by digital technology more generally.

Here, it becomes clear how these infrastructural geographies are implicated in the recomposition and multiplication of living labor. The role of computer-based tasks performed from home through crowdworking platforms in making accessible new digital wage workers (e.g., people with responsibilities for caring for dependents) or the complex spatiality of online games producing a curious form of digital racialization of labor and virtual migration are expressions of this. The focus on migration, gender, and other new and old forms of (spatial) stratification and fragmentation of labor is methodologically vital, particularly in a field in which such categories are often considered outdated in a global and digital world.

Digital infrastructure thus profoundly reorganizes the spatiality of labor from the macro level of the workplace to the global dimension, making new labor resources accessible while reorganizing old ones. This entails numerous consequences for the organization and composition of labor as well as for labor struggles, while also reconfiguring mobility practices and the gendered division of labor. The reconfiguration of labor through the digital factory is, crucially, a process of spatial reorganization; at the moment when the seemingly self-evident spatial architecture of the factory is called into question, the *spatial composition of class* might become central as Alberto Toscano has proposed.[10]

As I argue herein, the digital factory can take different forms. It can look quite similar to old industrial factories, but it might also be a digital platform or a video game. Despite this spatial and material variability, the digital factory has a great deal in common with industrial factories: They are infrastructures of production, characterized by the various types of technology that organize the production process, the division as well as the control and disciplining of living labor (often to the very last detail). Even if in the digital factory the elements and workers of a particular process of production might not be always assembled not under the roof of one building, digital technology, infrastructure, and logistics often allow for greater coherence and precision in the interplay and division of human labor and technology than in traditional industrial factory buildings.

RESEARCHING THE DIGITAL FACTORY

This book investigates several instantiations of the digital factory, ranging from Amazon warehouses to online video games, from gig economy platforms to data centers, from content moderation enterprises to social networks. They are all sites where digital technology produces labor relations, which allow testing the aforementioned concepts. In their peculiar ways, these sites are also important focal points, or prisms, to understand the current transformation of capitalism. The case studies presented herein involve digitized logistics in its increasing significance for global value chains, online multiplayer games not only as important sites of digital labor but also implicated in specific economic geographies generating previously unimagined forms of labor mobility, the importance of digital platforms in and beyond the gig economy, and the way labor is hidden in social media and digital infrastructure; arguably, all are not only sites of a certain form of digital and digitized labor but also focal points of contemporary political economy. Their numerous differences notwithstanding, the sites are tied together in multiple ways: throughout the book, we repeatedly encounter similar contractual arrangements, a curious return of piece wages, labor management software operating according to similar parameters, similar questions of space and infrastructure in relation to digital economies, labor legislation, or infrastructure such as data centers—or even, sometimes, the same corporations. It is precisely this multisited approach that allows delineating a tendency of the transformation of labor in digital capitalism, not specific to just one sector.

The book is based on more than seven years of empirical research in various sites using a range of qualitative methods, including ethnography and interviews, combined with the analysis of a variety of other materials and infrastructural technologies. Interviews played a central role, especially the conversations with workers were at the heart of this research. For each of the four case studies, I interviewed different groups of workers: warehouse workers at different Amazon warehouses as well logistics workers at other sites such as the Berlin Airport, different groups of workers in the gaming sector and on crowdwork platforms as well as content moderators for social media, among others. As a complement to these interviews, I also spoke with various trade union secretaries and activists, managers, experts, and other relevant actors in the respective chapters. Most interviews took place in person; others were conducted via telephone or video or even only in text chat. Sometimes the informal conversations (e.g., at the fringes of strike meetings, in the chat of a video game) were even more productive and informative than formal interviews.

This already hints at the importance of ethnographic approaches to the inquiry: wherever possible, I tried to be present at the digital factories in question. This meant I engaged in traditional (offline) participant observation at warehouses, logistic parks, or the offices and studios where games are produced, as well as union meetings, strikes, and other site visits for the logistics chapter and a part of the gaming chapter (these sites were mostly located in eastern Germany). The ethnographic research for the later chapters moved increasingly online (and further shed light on the complex and multiscale geographies of the digital factory). I spent months in the online multiplayer game World of Warcraft participating in the interactions and economics of the game and observing the digital shadow economy of "gold farming." The social world and political economy of online gaming can hardly be analyzed without deep attention to the online space of these games and their specific forms of not only digital labor but also interaction and sociality. The chapter on crowdwork is also, to a large extent, based on online (auto-)ethnography. I registered on different platforms, became a crowdworker myself, and began performing a multitude of tasks to understand the logic, infrastructure, and labor process of these platforms over the course of several months. I looked for addresses and locations, spent hours categorizing pictures, trained speech recognition software, searched the web for phone numbers, categorized thousands of fashion items, transcribed video files, and many other similar tasks. Although the platforms are designed to prevent contact with coworkers, I found every platform to be surrounded by an environment of various online forums and other social media communities.

In all the chapters, particularly that on crowdwork, important nodes—and crucial research sites for me—are constituted not only by the games or platforms themselves, but also by this surrounding ecology of social media, blogs, and forums. Forums, mailing lists, and social media are not only sites of socialization but also sources of mutual aid, organization, and resistance by workers. As spaces of solidarity and conflict, these communications are accordingly of critical importance to my research approach, which sought to proceed from (more or less visible) struggles and strive toward a better understanding of the conditions and new forms of strategy and organization necessitated by changing conditions. These forums were especially important for the analysis of online labor platforms in the chapter on crowdwork. In the world of online piece work, forums are spaces where workers meet and discuss a variety of issues, the analysis of which facilitated not only a deeper understanding of their social composition and subjectivity but also insights into the problematics of self-organization and resistance among the "crowdworking class."

The study of technology from small devices to huge infrastructures is another crucial component of the research underlying this book. Machinery has always been a major component of the organization and management of production in dictating the rhythm and way the labor is organized and performed; as such, it has also always been a site of contestation. In the age of ubiquitous digital computing and big data, however, the means of organizing, managing, measuring, and controlling labor and social cooperation are changing, as are the forms of conflict. Such development makes their analysis crucial. The attention to different infrastructures is part of a specific approach aimed to underline the materiality of digital capitalism. Unlike what terms like the "weightless" or "virtual" economy may suggest, the digital transformation of capitalism is in fact a deeply material process. Digital devices, satellites, fiber-optic cables, or data centers supplement older infrastructures such as harbors, roads, and railways.[11] Infrastructure includes, in my understanding, software. Software studies in recent years have underlined the importance of algorithms in terms of their far-reaching, complex, and contingent effects on the political, social, and material fabric of the contemporary.[12] The sites investigated in this book necessitate an understanding of the hard infrastructures of built spaces, the soft infrastructure of code, and the interplay of both. The study of such infrastructures and their role in production and circulation is crucial yet often difficult, as they tend to be hidden from view in various ways, often shielded by multiple layers of code and concrete.

This is particularly true of many of the software architectures used to organize, control, and measure labor, production, and circulation. Such software is usually opaque, sometimes even to the people using it, and researching it is a challenge. Throughout the research underlying this book, I addressed this problem by trying to grasp the functioning of these digital infrastructures initially through interviews with various user groups involved with them. In many cases, the workers who face the algorithmic and infrastructural architectures that govern their workdays are the best experts in deciphering their functioning. Ethnographic methods—that is, navigating these algorithms by, for example, working on online platforms or gaming myself—proved valuable to at least allow for a degree of speculation concerning the encoded logic. This often involves a second important method in researching software architectures, namely, experimenting with algorithms and also sometimes trying to reverse-engineer them (sometimes simply by playing with them and testing reactions to different inputs). It should be emphasized that these, as well as the other forms of research into software employed here, are limited by my rather small expertise in the world of coding.[13]

An analysis of a range of supplementary materials (e.g., union publications, press releases, economic reports, labor contracts, lawsuits) proved fruitful in understanding the functioning of the various digital factories in question. The sheltered operations of crowdwork platforms or content moderation providers are sometimes brought to light by lawsuits and whistle-blowers as well by the work of investigative journalists and researchers. The latter was an important source for the research, especially for the hidden world of content moderation for social media platforms. Another surprisingly helpful source of insights into otherwise sheltered algorithms and other infrastructures are patents. Corporations like Amazon make few public statements about the software used in their distribution centers or their e-commerce platform but end up saying a great deal in the numerous patents they file. An analysis of patents filed by Amazon and other companies provided another source to gather information about not only software but also "hard" infrastructures, which proved helpful in understanding, for example, the rationality of labor in the distribution center as well as Amazon's operations across different territories.

This methodology is clearly experimental in that it brings together very different research sites and methods to undertake a broad analysis and conceptualization of dynamic processes of global transformation. I am convinced that it is necessary to take up the challenge of carefully researching concrete sites of labor and struggle and, at the same time, try to situate the analysis of these places within a broader understanding of the transformation of global capitalism. In her essay "Supply Chains and the Human Condition," anthropologist Anna Tsing asks: "how can we imagine the 'bigness' of global capitalism (that is, both its generality and its scale) without abandoning attention to its heterogeneity?" She proposes the concept of "supply chain capitalism" to theorize "both the continent-crossing scale and the constitutive diversity of contemporary global capitalism" and "the structural role of difference in the mobilization of capital, labor, and resources."[14] Acknowledging the structural importance of difference and the importance of value as a crucial operator of these differences, David Harvey speaks of capitalism as "the factory of fragmentation."[15] An understanding of capitalism as a totalizing (but not total) mode of production that continuously broadens its reach is crucial to this book and informs my research, which in turn can contribute to a better theoretical and empirical understanding of the contemporary transformation of capitalism.

Searching for the digital factory does not mean ignore either the changes of recent decades (e.g., the real decline and relocation of many big industrial factories, especially in the Global North) or the shortcomings of factory-centered theory and politics (as outlined particularly but not exclusively by feminist approaches). I use the term *digital factory* to seek out unexpected continuities and new constellations and to understand the deeply heterogeneous effects of digital technology on the world of labor. As with any theory and research methodology, this book provides a necessarily incomplete picture and perspective—a perspective that is nonetheless aimed to be valuable to many people.

Like any study, again, this work does not start from scratch but builds on rich traditions of empirical and theoretical research, to which it is greatly indebted. A crucial starting point and important source are the insights of Italian workerist Marxism and its attempts to come to terms with the transformations of social production (not only) in postwar Italy from the 1960s up to the present, with its dynamic understanding of technological and political change driven by social struggles. More recently, writers from this tradition have engendered discussion around the transformation of labor in post-Fordism, most notably with the important concept of "immaterial labor."[16] This term must be understood in the context of postworkerist ideas such as "mass intellectuality" and a renewed interest in Marx's idea of the "general intellect."[17] All these ideas refer to new forms of creativity and communication involved in social production that has reached new levels of cooperation, which increasingly operate independent from the command of capital. As Antonio Negri argues, today "we face a new technical composition of work: immaterial and service-based, cognitive and cooperative, autonomous and self-valorizing."[18]

The rich debates that have evolved around the idea of immaterial labor and its theoretical ecology are in many aspects the central starting point of this book—a starting point, however, from which it quickly departs. The ideal immaterial workers of this debate are more often than not highly qualified but precarious urban workers of the Global North characterized by their communicative and creative labor (e.g., designers, programmers), who operate at a distance from capital's direct command. In a sense, this book addresses the opposite type of worker. Exemplified by the term *digital Taylorism*, it foregrounds other qualities and sectors of labor in digital times. My research is interested in how digital technology (in its most advanced forms) brings forth labor relations characterized by, for example, tight control over the entire labor process as well as rationalization and decomposition. With that, the focus moves to the other groups of workers; this is intended not so much as a critique of theories of immaterial labor as a supplement and moving of focus and perspective.

Hereby, my research builds on the work of authors such as Nick Dyer-Witheford, George Caffentzis, Ursula Huws, and Lilly Irani, who put the spotlight on sites where the results of digitization bring forth labor conditions that are not marked by free communication and creativity but by deskilling, routinization, and control.[19] Behind these more recent approaches, of course, a diverse and rich predigital history of critical investigations into the relation of technology and capitalism constitutes an important, although possibly less visible, source for this book. These traditions and debates include, for example, the early writings of Italian workerism, in which the concept of the "social factory" provides a brilliant early example of thinking of the factory beyond its walls or, of course, Harry Braverman's trenchant critique of Taylorism.[20]

My work ties in with such approaches to show the necessarily heterogeneous composition of labor in the digital age and to shed light on one tendency of development that is theorized as digital Taylorism. My claim is then that digital technology, especially in its most advanced forms, creates a set of very different labor situations, in which a new digital Taylorism exists alongside more autonomous forms of (immaterial) labor. In describing digital Taylorism as an important sector, I am not claiming it to be the new hegemonic form of labor but, rather, one tendency among others, albeit one that has recently increased in importance. Thus, digital Taylorism exists alongside other—markedly distinct—labor regimes. This coexistence alongside other labor regimes is not circumstantial but in fact necessary. The existence of heterogeneous, variegated, and fragmented labor regimes is then neither accidental nor a simple question of uneven development but, rather, an important characteristic in the development of capitalism, which has, if anything, grown in importance over time.

THE BOOK

The book consists of this introduction, four chapters focused on case studies, and a concluding one. The following chapter, "The Global

Factory" moves the spotlight to the field of logistics. It starts with the shipping container and the algorithm as crucial elements of a development that has shifted logistics to the center of global capitalism since the 1960s. The first focus of the chapter is an investigation into labor and struggle in Amazon distribution centers, and one facility near Berlin in particular. The distribution center is a focal point of the insertion of human labor into circulatory systems on a grand scale, increasingly organized by algorithmic architectures. Distribution centers are crucial sites of logistical labor governed by software, with regard to both minute details as well as the level of global supply chains. At the same time, Amazon's distribution centers have been sites of a prolonged labor conflict over these conditions for several years. The second part of the chapter moves the focus to the next part of the supply chain: delivery on the *last mile*. With the rise of online shopping, the last mile has become a crucial point of logistical operations in urban contexts, and a site of intensive competition and experimentation with hyperflexible labor. Forms of app-based labor further radicalize the flexibility of logistical labor and allow to situate logistics as an important element in the genealogy of the gig economy.

The following chapter, "The Factory of Play," addresses labor and the political economy of video games. Two investigations focus on digital workers in the German gaming industry and so-called gold farmers, professional video game players located primarily in China. The German case focuses on instances of struggle and unionization on the part of gaming workers in Berlin and Hamburg. At the heart of the investigation are not the celebrated game designers and artists but a team of testers working in quality assurance (QA), whose labor consists of playing the game in order to find errors. Even more so than in the Berlin-based QA department, the labor of Chinese "gold farmers" is repetitive, exhausting, and characterized by long hours and tight workplace discipline. Hereby, professional players (gold farmers) earn in-game items to sell to predominantly Western players who want to advance in the game. The result is a complex economic geography in which the Chinese digital workers become virtual migrants to Western servers, where they work in a digital service sector while facing racist abuse. The chapter analyses the peculiar nature of this digital shadow economy and its reconfiguring of migrant labor and racism as part of the political economy of online games and the specific forms of labor and circulation it produces.

The next chapter, "The Distributed Factory," is concerned with crowdwork. Crowdwork platforms are digital platforms that allocate tasks to a global pool of digital workers, most of whom work from home on their personal computers. These digital home workers are a crucial but hidden component in the production and training of AI. Work on these platforms is characterized by decomposition, standardization, automated management, and surveillance, as well as hyperflexible contractual arrangements. The chapter investigates the composition of the crowdworkers themselves, putting forward the argument that the organization of work described in the first section is precisely what allows for the multiplication of labor. The platform as a digital factory is capable of assembling a deeply heterogeneous set of workers while bypassing the need to spatially and subjectively homogenize them. As a result, crowdwork taps into labor pools hitherto almost inaccessible to wage labor, such as women shouldering care responsibilities who now can work on crowdwork platforms while at the same time performing domestic labor. The extension of mobile internet infrastructure in the Global South in recent years also translates into access to a massive pool of potential digital workers.

The following chapter, "The Hidden Factory," situates labor's place in social media. The chapter turns to aspects of the political economy of social media that are often obscured behind debates about hate speech, privacy, and data protection—namely, the infrastructure (both hard and soft) of social media and the labor hidden within this infrastructure. Exploring the labor behind code, data centers, and content moderation, the chapter sheds light on the often hidden forms of labor that are crucial to both social media and other sectors of the digital economy. Buried in these infrastructures we also find forms of labor such as content moderators or "raters"—human workers whose job is to improve search algorithms—workers very much akin to the yellow badges scanning books for Google.

The conclusion synthesizes the theoretical and empirical analysis, arguing that digital capitalism is not characterized by the end of the factory, but by its explosion, multiplication, spatial reconfiguration, and technological mutation into the digital factory. The digital factory takes very different forms and seldom looks like a traditional factory building. One such configuration that is increasingly central is that of the platform, which we encounter in very different forms throughout the book. The platform might be the paradigmatic factory of today's digital capitalism. Thereafter, an epilogue evaluates the impact of the COVID-19 pandemic on the developments described in the previous chapters.

Taken together, these chapters provide a picture of the digital factory as a flexible and mutable form: a distribution center, a video game, an internet café, a digital platform, the living room of a digital homeworker. These factories are populated by very different and heterogeneous groups of workers often distributed across different geographies. Nevertheless, the digital factory is an infrastructure able to synchronize these workers into algorithmically organized production regimes. It is here that we find its characteristics, across all differences and locations, have qualities in common with how labor is organized, distributed, divided, controlled, and reproduced. It is here that we see the contours of a labor regime that is a crucial component of digital capitalism in the very process of restructuring the social division of labor, its geographies, stratifications, and lines of struggle.

2

THE GLOBAL FACTORY

Logistics

Some time ago, the American sociologist Thomas Reifer speculated in a lecture that if Karl Marx were to write his most important work nowadays, the opening words would be different. As is well known, the famous first sentence of the first volume of *Das Kapital* reads: "The wealth of societies appears as an 'immense collection of commodities.'"[1] Today, suggests Reifer, this wealth appears far more as an "immense collection of containers."[2] For Marx, commodities are not simply things; similarly, the container is not simply a box used to store things. In its material and symbolic functions, the container refers to a contemporary capitalism in which the movement of goods plays an increasingly central role. It is a symbol and precondition of a profound logistical transformation of global capitalism. As such, the container is a paradigmatic expression of a world set in motion through logistical operations and infrastructures, and a symbol for the rise of this discipline.

In attempts to represent this transformation, the container has also become a central element, an icon of globalization, a ubiquitous symbol of global trade, and the object of artistic attempts to depict today's global capitalism. In their winter 2013 mission, a self-described nomadic design studio calling itself Unknown Fields Division traveled to the South Chinese Sea aboard a container ship. Titled *A World Adrift*, the investigation takes ships and containers as points of departure for an analysis of contemporary capitalism's spatial and territorial qualities. Following the Gunhilde Maersk, a container ship with a capacity of over ten thousand standard containers, from the coast of southern China to the UK, they show the central role of these steel boxes for the circulation of goods. Endless rows of containers stacked against the backdrop of smoking factory chimneys are traversed by cranes that load and unload the containers, seemingly without any human interference. These are logistical landscapes, landscapes characterized by steel and standardization, the rhythms of huge machinery, and the dwarfing of human labor. These landscapes constitute "'really abstract' spaces," which seemingly make the abstract logic of capital representable.[3]

Although Unknown Fields Division is clearly fascinated by the never-ending flow of containers, the studio is careful to avoid taking the mechanical landscapes of endless containers as the whole picture. The movement of containers is complemented by impressions from the factories of southern China, producing the goods to be exported. Similar to artist and filmmaker Allan Sekula, another chronicler of the container, the studio breaks with the homogeneous visions of the logistical landscape in the attention paid to the margins of logistical architectures and their uneasy relation to neighboring spaces and populations. Unknown Fields Division visits the South China Mall in Dongguan, which, as the world's largest shopping center, hints at the fact that the flow of commodities from East to West is not a natural fact; in addition, the mall's desertion and decay mirror some of the contradictions of Chinese state capitalism. The perhaps crucial dimension of these investigations is the ongoing centrality of human labor in the face of gigantic technological infrastructures. Both Sekula and Unknown Fields foreground the central role living labor continues to play in production as well as in circulatory systems. The workers in China's factories and harbors and the workers aboard the container ships are all represented as both key components of the circulatory systems and possible actors in their future interruption.

This chapter explores both the rise of logistics and the persistence and transformation of logistical labor against the backdrop of container- and algorithm-driven automation. While I am very much interested in the logic of standardization and abstraction embodied by container and code, this chapter is committed to a perspective that resists the picture of a seamless logistical harmony of endless circulation. The effect of logistics is a process not of smooth globalization and the disappearance of borders but, rather, of their multiplication and flexibilization—just as logistics does not mean the end of labor but its displacement, multiplication, and flexibilization. This argument is central to not only this chapter but the entire book. It hints at the fact that many of the logics involved in containerization can also be found in the manifold processes of digitalization; accordingly, as technologies of abstraction, container, and code exhibit similarities not only in their homogenizing drive but also in how they create heterogeneity and fragmentation. In many ways, this book addresses the processes through which the establishment of standardization, homogeneity, and intermodality—through container and code, software, and infrastructure—allows for the multiplication and heterogenization of space and labor.

Labor remains crucial to logistical operations. Despite all efforts at automation, it is still critically important in maintaining circulation. This chapter visits various sites of logistical labor, most importantly, the distribution center as a crucial node in supply chains. Here, as in other sites of logistical labor, digital technology is transforming labor regimes and processes profoundly. Both digital Taylorism and the multiplication of labor prove to be crucial principles in the context of logistical infrastructures. Next, the chapter follows the goods leaving the distribution center and entering their "last mile" to the customer's doorstep. In recent years, the last mile of delivery has become a crucial focus of logistical operations in urban contexts due to the rise of online shopping and the spread of platforms ranging from Amazon to Foodora, Deliveroo, and others. The increasing importance and time sensitivity of delivery reconfigure both urban spaces and labor relations. Accordingly, labor in the last mile is characterized by intense time pressure, standardization, algorithmic management, and digitally enabled surveillance on the one hand and platform-driven precarization and flexibilization on the other.

Before moving on to the detailed study of labor in logistics, however, it is necessary to start with the rise of logistics as both an industry and a rationality by sketching out a historical process, often described as the "logistics revolution." This process involved a transformation of not only an industry but, more importantly, capitalism itself. In examining containers and algorithms as technologies essential to the rise of the logistics industry, this chapter starts by roughly tracing how logistics has become a crucial economic sector and how it permeates and fundamentally shapes today's capitalism.

THE CONTAINER, OR THE LOGISTICS REVOLUTION

April 26, 1956 is often cited as the starting date of container shipping. On this day, a refitted tanker sailed out of a New Jersey port. Aboard were fifty-eight steel boxes anchored to the deck that could be directly transferred to trucks by cranes. The system was developed by the transport entrepreneurs Malcolm McLean and Roy Fruehauf. Their innovation consisted of little more than stackable steel boxes with a twist-lock system to secure them, which could be transferred directly from trains or trucks to ships using specialized cranes. Moreover, it was not even particularly new; container systems and attempts toward standardization had existed for more than a century. It was, however, McLean's and Fruehauf's system that finally prevailed. The US military played an important role in this success, as they adapted the system to the logistical requirements of the Vietnam War. Between 1968 and 1970, four International Organization for Standardization (ISO) standards provided the foundation for the intermodal shipping containers in use today: The twenty-foot equivalent unit (TEU) describes the approximately twenty-foot-long standard container used to calculate loading and shipment volumes.

The intermodal container dramatically changed ports. The time and space needed to transship significantly decreased, and far less labor was required. Although some unions attempted to delay or regulate the process of containerization, they ultimately conceded defeat. In this sense, containerization also signaled a blow to the power of organized labor in the ports—traditionally militant and internationalist sectors of the union movement as a whole—such as the International Longshoremen's Union. The port as space, as a working place and transshipment point, changed quickly and dramatically as a result of containerization. At the same time, containers began to play an ever-greater role in the transport of goods. Today, almost all international trade revolves around the standardized container. Ninety percent of general cargo is transported in containers on ships, which in turn account for 90 percent of the global transport of goods. Major harbors (e.g., Hamburg, Rotterdam) can handle more than twenty-five thousand standard containers daily. Containerization also contributed to the rise of shipping companies. The Gunhilde Maersk, the container ship that hosted the Unknown Fields research trip mentioned in the introduction, is owned by Maersk Lines, one of the most important shipping companies in the world today. Although Maersk operates over six hundred vessels carrying 2.6 million TEU, has offices in over one hundred countries, and employs tens of thousands of sailors and other staff, it is (like many logistics enterprises) fairly unknown to the general public. This also illustrates that logistical operations and infrastructures are highly present across many spaces but are generally recognized only in when they fail, a circumstance which characterizes logistical operations and infrastructures as part of a "technological unconscious" of global capitalism, to borrow a concept from Nigel Thrift.[4]

The shipping container is the fundamental technological precondition and a symbol of globalization. It stands for the principle of standardization and modularization and has contributed to a massive increase and acceleration in global circulation. As such, it is also a prerequisite for the second globalization of the twentieth century. A focus on the container and the logistics revolution yields a different kind of historiography of this globalization. While most narratives of neoliberal globalization concentrate on free-trade agreements, structural adjustment programs, and, accordingly, such institutions as the World Bank and International Monetary Fund, this alternative narrative would place greater emphasis on such technologies as the container, on the rise of logistics, and as a result on such institutions as the aforementioned ISO and transnational shipping operations. Such a perspective makes it possible to retrace the material history of globalization and thus provides an important appendix to many standard narratives.

Its centrality notwithstanding, the container is only one piece of a variety of factors comprising the logistics revolution—"the most underinvestigated revolution of the twentieth century," as geographer Deborah Cowen argues.[5] While questions of transport, distribution, and storage are of elementary importance for virtually every type of economic activity, the term *logistics* is historically derived from not the civilian economy but the military. The movement of troops and materials over long distances, the transport of supplies, the control of streets and bridges—these are all decisive matters of warfare, in the context of which the term *logistics* initially emerged in the nineteenth century or even earlier. Another origin of logistics, in this case civilian, is the postal system. Here, the problem of measuring or mapping of space for the purposes of transport came to light quite early. Stefano Harney and Fred Moten posit yet another genealogy of logistics, noting its origins in the transatlantic slave trade.[6]

The more recent economic origins of logistics as we now know it can be traced back to the so-called logistics revolution that took place in the 1950s and 1960s, mainly in the US. Here, the term migrated from the realm of the military to the civilian economy—in no small part because many logistics experts in the military undertook this same migration after World War II. While up until this point transport and storage were generally seen as steps to be completed as inexpensively as possible following production, under the label "logistics" a new organization of the entire supply chain was conceived, including everything from design, ordering, production, transport, and warehousing to sales, modifications, and reordering. Logistics increasingly defined the entire cycle of production and distribution as something to be planned and analyzed. This shift in perspective elicited the principles of modern logistics and set the changes in motion now subsumed under the term "logistics revolution."[7]

This change in perspective began to gain momentum in the 1950s and 1960s, when logistics emerged as a management paradigm, became a subject of academic study, and was connected for the first time to digital computing systems. Logistics became an increasingly central aspect of economic planning, a field of knowledge, a rationale. The shift from transportation to logistics, "from *practical afterthought* to the *calculative practice that defines thought*" describes how the physical circulation of commodities grows in strategic importance for capital.[8] Rather than transportation as a necessity following production, logistics came to be understood as a crucial element in the production of surplus value.[9] The further integration of production, circulation, and, increasingly, consumption can be identified as a central effect of the logistics revolution—a tendency that, incidentally, is evident in Marx's scattered remarks on the "physical conditions of exchange."[10] Clearly, production has always included steps of transportation—think of the assembly line—but today's logistics engenders a tendency for commodities to be "manufactured *across logistics space,* rather than in a singular place," to use again the pregnant words of Deborah Cowen.[11] This hints at the growing significance of physical circulation, which goes hand-in-hand with an increasingly blurry distinction between production and circulation.

THE ALGORITHM, OR THE SECOND REVOLUTION

If containerization can be understood as a crucial element in the rise of logistics, the rise of digital technology is without a doubt another such element. Put simply, one could argue that the computerization of logistics entails a second logistics revolution, once again changing the industry and the global circulation of goods profoundly. In its logic, digital computing is in a certain sense quite similar to the container: standardization, modularization, processing. In the realm of logistics, digital technology has spread at an uneven rate and to varying degrees, but it now pervades most logistical operations and has contributed to the further acceleration of global circulation. This ranges from the tracking of individual products and the detailed control of individual working processes in a distribution center to highly complex software architectures that oversee, analyze, and coordinate entire supply chains. Digital technologies, with their special infrastructures, have social, spatial, and political implications. In the field of logistics, they serve primarily to organize, measure, control, and predict the movement of goods (and humans). Increasingly interconnected across the entire supply chain and at the same time increasingly autonomous, these systems generate effects that humans are less and less able to oversee.

The digitization of logistics includes a multitude of dimensions—for example, shipping software, enterprise resource planning (ERP) systems, global positioning systems (GPS), barcodes, and later radio frequency identification (RFID) technology, and the corresponding infrastructures. All these technologies to organize, capture, and control the movement of people, finance, and things are described by media theorist Ned Rossiter as "logistical media."[12] Logistical media compromises both small devices and increasingly complex software architectures managing logistical operations. An example of the importance of such algorithmic governance is ERP software, digital real-time platforms geared toward the integration of all parts of a company (e.g., financial management, logistics, sales and distribution, human resources, materials management, workflow planning) into one program. Generally speaking, ERP software is expensive proprietary software produced by a small number of companies. One of the most important players is the German firm SAP, which claims to deliver its manifold software products to 87 percent of the *Forbes* Global 2000 corporations.[13] Martin Campbell-Kelly argues that the importance of SAP's software in relation to its prominence is inversely proportional to that of Microsoft's product, and speculates that should SAP's ERP system cease to exist, it would take years for substitutes to take over and close the breach in the global economy; in contrast, Microsoft's broadly used software could be substituted within days.[14]

Often, ERP systems and similar software are opaque in their functioning, even to the individuals operating them. Although these software architectures are at times incredibly complex and able to deal with a high degree of contingency, they adhere to a specific logic that tends toward abstraction and standardization. Hence, protocols, parameters, standards, norms, and benchmarks are key to the organization of goods and labor by semiautomated management systems. Designed to monitor, measure, and optimize labor productivity and supply chain operations, logistical media are crucial to organize logistical operations along different supply chains. These forms of logistical media are also an important component in the organization of labor in logistics.[15] As this chapter elaborates, digital technology is crucial to the insertion of living labor into logistical operations. At the same time, the standardization and real-time surveillance of labor processes are precisely what allows for the flexibilization and multiplication of labor in the sector.

Like containers, algorithms play an essential role in a massive acceleration and the ever-increasing importance of logistics operations. They are also technologies of standardization and modularization that have contributed to the rise of logistics as a central discipline in modern-day capitalism. Container and algorithm are essential infrastructures of globalization, responsible for wide-ranging economic changes. A crucial impact of the logistics revolution driven by containers and algorithms is a shift in power between companies that produce goods and those that sell them. The rise of the retail giants Amazon and Walmart, two of the biggest and most important corporations of today's economy (this history would need to be completed by the rise of Alibaba, the Chinese retail giant that has emerged as the most important global competitor for Amazon), is closely tied to the logistics revolution and expresses this power shift.

THE RISE OF RETAIL

The rise of the retail giant Walmart to the world's largest revenue-generating company is closely tied to the shipping container. The company's low-price/high-volume business model is based on the import of mass-produced goods. Annually it imports some seven hundred thousand standard containers to the United States alone. Ostensibly a retail operation, Walmart is essentially a logistics company that employs a strategy based on effective just-in-time logistics, the minimization of storage space, and the precise forecasting of customer behavior based on huge volumes of data. Like other large retailers, Walmart strives to control the entire supply chain and is in a position to dictate to most manufacturers the conditions of production and acquisition. The rise of the company can be attributed to a logistics strategy and is the expression of a shift caused by the logistics revolution. While production and distribution are increasingly being merged, it is often large retailers like Walmart or Amazon that dominate supply chains and dictate conditions to manufacturers. Walmart's strategy in terms of not only of global supply chains but also the geographical allocation of its stores and distribution centers and its innovative sales architecture is demonstrated simply but pointedly by Deborah Cowen's characterization of the logistics revolution as "a revolution in the calculation and organization of economic space."[16]

The company's spatial planning revolves around its distribution centers, tight control of the entire supply chain, and innovative computerized inventory management as well as the precise forecasting of customer behavior based on huge volumes of data warehoused in the company's own data centers and analyzed by over two thousand data experts hired to predict and model customers' desires and preferences. Everything is designed to accelerate the turnover of goods and minimize storage costs—crucial factors that have made Walmart by far the world's largest company in terms of revenue. Walmart's logistical network is driven by a huge digital infrastructure that generates and processes high volumes of data, which are used to organizes processes and to increase efficiency and make forecasts. A local Walmart hub, for example, casts a seventy-terabyte information shadow.[17] Walmart recognized early on the central role of digital information and communication for modern just-in-time logistics. As early as 1987, it created its own network of satellites to the tune of $24 million—at the time, the largest private network worldwide.[18]

In contrast, the rise of Amazon (whose founder, Jeff Bezos, took Walmart as inspiration) is the story of the rise of e-commerce. In terms of both revenue and market capitalization, Amazon is currently one of the largest corporations worldwide. Started as an online bookstore, Amazon today offers a wide range of further services and products. Its Amazon Web Services (AWS) branch, for example, is one of the most important owners of data centers and cloud computing infrastructure on the global market. Even if AWS attracts little public interest, it has made Amazon the most important provider of cloud computing services on a global scale, with customers ranging from the streaming platform Netflix to the Central Intelligence Agency. Amazon commands a highly profitable infrastructure including proprietary software, data centers, and even high-speed undersea cables.

Amazon offers a wide range of other services and products and has even begun selling its own clothing and consumer products under the Amazon Basics label. It opened its first brick-and-mortar "smart" stores tied to its launch of Amazon Go technology, which uses sensors to automatically charge customers when they leave the store and thus does not require cashiers. Amazon also produces consumer electronics like the Kindle e-book reader, a tablet, and the smart home application Echo. With Alexa/Echo, Amazon stands at the forefront of consumer-facing artificial intelligence, a sector with the potential to become another central pillar of the company's growing empire. As in other departments, the massive amount of data Amazon can mine proves central to its business and its expansion strategy.

Amazon has also emerged as a producer of movies and television series with its Prime Video department, whose productions have won several Academy Awards and Golden Globes. These operations constitute a closely intertwined ecosystem of media, devices, content, and applications centered on its ubiquitous shopping platform, which undeniably remains the core of its business model. Speaking at a conference on the in-house production of Amazon Prime movies and television series, Bezos offered a glimpse into the strategic thinking behind this ecology, explaining, "we get to monetize this content in a very unusual way. When we win a Golden Globe it helps us sell more shoes."[19] What he meant is that consumers who subscribe to Amazon Prime also purchase more on the platform, while successful movies and series help attract more people to Prime and retain their subscriptions. (Over one hundred million people in the US alone are Prime members.) This illustrates how most of its business model continues to revolve around the e-commerce platform. The platform's strength derives from its huge assortment of products, including almost all thinkable commodities, hundreds of millions of which can be ordered online. Behind the platform lies a massive logistical infrastructure that allows the goods to reach the customers at an ever-increasing volume and velocity. In this infrastructure, the company's infamous distribution centers represent a crucial pillar.

CHRISTMAS FEVER: INTO THE DISTRIBUTION CENTER

December is a special time for Amazon's distribution centers. These huge warehouses, known as "fulfillment centers" (FCs) in company lingo, experience their busiest phase of the year. Products are stored in these distribution centers, and after a sale is made on the website, these products are prepared for the shipping to the customer as fast as possible—a major factor in the competition between online commerce providers and offline shopping. Christmas season is the most important month for online sales, as for many brick-and-mortar shopping malls. The 2014 Christmas season was particularly uneasy at the Brieselang FC. Brieselang is a small city close to Berlin, the capital of Germany, Amazon's most important market after the US. Opened in 2013, Brieselang FC is one of an increasing number of German distribution centers run by Amazon. It is located in an area close enough to Berlin to dispatch deliveries quickly, but far enough away to guarantee the kinds of cheap rents and low wages common in rural eastern Germany. In December 2014, most workers were employed on temporary contracts: roughly 300 permanent employees shared space in the distribution center with more than 1,200 temporary workers, many of whom were hired only for the challenges of the Christmas season. In many distribution centers, the workforce doubles in the month before Christmas. A year earlier, on a memorable day a week away from Christmas, Amazon customers in Germany placed a record 5.3 orders per second, or a total of 4.6 million products in one day.[20]

These were the challenges faced by the distribution center's workforce, which was divided into holiday helpers, a few regular workers, and the other temporary workers who had fixed-term contracts that were renewed every six months. In Germany, this practice is only allowed for up to two years, after which labor law requires Amazon to either hire them permanently or let them go. Surprising most workers, Amazon management began informing the first group of employees about their future at the FC on December 22. For those who started working two years prior, only two options remained: either a permanent contract or searching for a new job. By the end of December, it became clear that the latter would be the case for most. Only thirty-five workers were given permanent contracts, many received new temporary contracts lasting between one and six months, but the rest were told to leave the FC immediately. "After the break, they started to inform small groups of about fifteen to twenty of us that we have to go immediately. They had also increased security personnel, we couldn't even say goodbye to our colleagues," remembers one of the workers whose contracts were not renewed.[21]

The mood had grown tense at the FC in the weeks leading up to Christmas. Rumors circulated about the possible closing of the Brieselang location, further fueled by Amazon's recently opened new distribution centers serving the German market in Poland and the

Czech Republic. Many workers and unionists felt these new Amazon FCs were also a reaction to the ongoing strikes in the company's German facilities. Since 1,100 workers had participated in the first strike in Amazon's largest German FC, Bad Hersfeld, in April 2013, the logistics giant had been hit by a years-long wave of strikes at various locations in a tenacious conflict with the trade union ver.di (United Service Sector Union) over collective bargaining, higher wages, and better working conditions.

Brieselang had not witnessed any strikes. Although the first works council election held over the previous summer had been a small success for ver.di, workers organized in the union felt that Brieselang was not ready for a strike. Workers with temporary contracts were especially hesitant to strike, fearing it could diminish their chances for a new contract. As if to bolster this assumption, FC management refused to renew the contracts of workers active in ver.di and the works council. "I feel punished for my perfectly legal participation in the works council. I think Amazon wants to create a climate of fear," said one, who went to court over the nonrenewal of his contract and would lose the case six months later.[22] Since the beginning of the dispute, Amazon has proven itself a tough antiunion employer, in stark contrast to the more cooperative approach many German corporations take. From attempts to discourage worker activity and workplace organization at the FCs to the new distribution centers in Poland and the Czech Republic and carefully launched press releases announcing progress in labor automation at the FCs, Amazon took an aggressive stance against concessions to the union and striking workers.

This ongoing conflict serves as the backdrop for my investigation into labor in Amazon's FCs. These distribution centers are crucial sites of logistical systems that remain highly dependent on human labor. The organization and government of this labor serve as a first instantiation of digital Taylorism. In the following chapters, the sites of labor will change, resembling the traditional factory much less than Amazon's distribution centers, which in comparative terms still exhibit many similarities. That said, many qualities of the labor process—its mobilization, flexibilization, and insertion into globalized and digitized circulatory systems—remain surprisingly similar. In the sense that an FC's spatial configuration is somewhat similar to the factories which birthed Taylorism, they may also serve as an apt starting point for a discussion of digital Taylorism—even if they arguably have as much in common with a crowdworking platform as with a factory of the early twentieth century.

The Brieselang FC exhibits many similarities to Amazon distribution centers around the globe. They tend to be located close to urban centers at sites with well-developed transportation infrastructure (in this case, roughly thirty kilometers from Berlin), although the areas themselves are often plagued by high unemployment and low economic development. Beyond the obvious advantages, these regions provide in the form of a pool of low-wage labor and cheap rents, local governments are often willing to grant huge subsidies and infrastructural development to attract Amazon's business.

The gray factory-like building in Brieselang is divided into six smaller halls spanning an area of sixty-five thousand square meters—or ten football fields, Amazon's standard unit of measurement for its distribution centers (the biggest FC in Phoenix, Arizona, stores fifteen million items in an area the size of twenty-eight football fields). A huge poster or print hangs in most FCs, bearing the slogan "Work Hard. Have Fun. Make History." This is only one of many features shared by most FCs. The distribution centers normally employ between one thousand and four thousand workers in a shift system varying according to local conditions and regulations. In Germany, FCs are typically active between 5:30 and 23:30; night shifts are rare in most locations, in contrast to Amazon's UK facilities, where they are part of the standard operating procedure.

Most workers in Brieselang come from the surrounding area, a region known for its high unemployment rates. Most had already worked in a variety of professions before coming to Amazon and were often unemployed for a longer period of time before being sent to the FC from the local job center. In this case, not showing up for work results in deductions from unemployment benefits. Reports from various distribution centers alleged that Amazon systematically exploits the job center subsidies designed to reintegrate the unemployed back into working life by laying recipients off after the subsidies expire. The local workers are complemented by a small group of migrant workers, some of whom are from Spain and Turkey, and a small segment even makes a multihour daily commute from neighboring Poland. Most FCs initially draw their workers from surrounding areas. Older distribution centers illustrate how this circle expands over time, as most FCs exhibit high worker turnover rates, and Amazon has sometimes even been forced to implement a bus system to shuttle workers from as far as seventy kilometers away.

Each shift at Brieselang begins with a short motivational speech by the supervisor, called "leads" or, one step up on the FC hierarchy, "area managers" in the company lingo. These talks focus almost exclusively on performance and goals. Every item that enters the FC, changes position, or leaves is precisely registered, and every team of workers is given a daily quota to be reached. Before workers can take their positions, they are forced to "badge in" and pass airport-like security scanners. Every movement throughout the distribution center is regulated and standardized—signs instruct workers to use the handrails, and yellow markings indicate the correct walking paths through the FC. Many workers feel patronized by the exacting protocols they are obligated to follow. From Amazon's perspective, however, standardization is a crucial element of its business model, and the processes in the FC are optimized to the finest details.

WORKING TO THE RHYTHM OF BARCODE AND SCANNER

Brieselang's exterior is characterized by the red and yellow trucks and containers of the logistics company DHL, which bring goods to the distribution center and pick them up for delivery to customers. Amazon relies on various companies in Germany, but DHL is by far the most important. It was bought by the formerly state-owned German postal monopoly Deutsche Post in 2002 and is among the most significant logistics corporations globally. One of its central cargo airports, Halle-Leipzig, is close to Brieselang, as is the Port of Hamburg, one of the most important European container ports. Experts estimate that every seventh parcel in Germany is sent by Amazon, making DHL and other logistics providers extremely dependent on good relations with the company—a fact used to exert pressure on providers in terms of price and priority.

After incoming goods are unloaded into the distribution center from the back of the trucks, the packages are scanned for the first time and opened at the "Inbound Dock." Assuming they are standard sized, they begin their movement through the FC on one of the many conveyor belts stretching like a web throughout the entire FC. The next station is "Receive," where workers separate the goods into single units, check if they are broken, and scan them again. Amazon requires its suppliers and third-party merchants to deliver the goods with barcodes on both the delivery boxes and the individual items. All distribution centers are governed by the logic of the barcode: "everything at the FC has a barcode, even me," explains one worker, "and the beeping of the scanner is the sound of my work."[23]

At 8:01 on June 26, 1974, cashier Sharon Buchanan scanned a ten-pack of Wrigley's Juicy Fruit chewing gum at the Marsh supermarket in Troy, Ohio. This marked the first commercial application of the UPC barcode, which still serves as the basis for most of today's models.[24] Based on Morse code, the system had existed since the 1940s but only became effective with the invention of laser scanners like the one used by Buchanan. Again, the crucial point was not so much the invention of the technology itself but its standardization across producers and retailers. Alan Haberman, former chair of the Barcode Selection Committee, an industry committee that clashed over which system would become the standard, commented on their efforts: "This little footprint . . . has built a gigantic structure of improvements of size and speed, of service, of less waste, of increased efficiency. This lousy little footprint is like the tip of an inverted pyramid, and everything spreads out from it."[25] Today, the barcode is used in multiple ways to recognize goods in shipping, postage, and personal shopping by millions of companies that scan billions of barcodes each day. Designed for machines rather than the human eye, it is a clear example of the power of code and standardization. Amazon's distribution centers demonstrate that the barcode is not only a technology of recognition but also an example of logistical media technology, insofar as it facilitates the following of objects (and workers, as discussed later) through time and space.

At any distribution center, every good is scanned and entered into Amazon's inventory database at the receiving station. Goods arriving at any distribution center for the first time are also precisely measured and weighed before the results are shared globally among all distribution centers. After scanning, the worker on the receiving end puts the goods into "totes," as the omnipresent yellow plastic bins are called in company lingo. These boxes form the basic logistical unit of the distribution center, which are then placed back onto the conveyor belt to move on to the next station, called "Stow," where workers put the boxes in carts and sort them into shelves across the distribution center.

The storage system used in the FCs follows no logic apparent to the bare eye. Rather than being sorted into designated sections, goods are placed seemingly at random onto unoccupied shelf positions. Whereas systematic storage systems allocate the same goods and classes of good to designated areas (e.g., one area for books, another for toys), at a distribution center using random storage there is no such obvious logic to the system, with books next to iPads, toothbrushes, and toy cars. Random storage is particularly attractive for companies like Amazon, which carry rather small stocks of a wide range of products and orders tend to combine products from different categories. Apart from these factors, general advantages of random storage include more efficient use of free space, flexibility concerning changes in product range, and fewer picking errors as the products next to each other tend to be different.

Random storage does not rely on human agents knowing the position of inventory, but on software. Only Amazon's software knows exactly where each good is stored and can guide the workers to it. In cases of software problems or power outages, random storage instantly becomes a chaotic mess where nothing can be found. When it works, however, random storage can optimize both the storage capacity and overall efficiency of warehouse operations. At the Brieselang distribution center, each shelf position carries a unique barcode that is scanned by the "stower" using a handheld scanner together with the product barcode. From this point onward, the software knows exactly where a product is stored. Like almost all other goods in the FC, it is now ready to be picked as soon as it catches the eye of an online shopper in the FC's vicinity.

A product may sit on its shelf for minutes, days, months, or years, listed on Amazon's website and ready to be sold. After a customer decides to buy a product via Amazon's home page, the order is routed to the appropriate FC by Amazon's software where, in most cases, it will take only hours for the distribution center's next group of workers to be set into motion: the pickers. Workers in the FCs can sometimes guess the weather outside based on the intensity of orders, as orders tend to go up when it rains. Like the workers in stow, the picker's equipment consists of a handheld scanner and a trolley with which they navigate the shelves in the picking tower—movement through the space of the FC is dictated by the commands of this scanner. It provides the location of the item to be picked, which must be scanned once the shelf is reached. The picker takes the item, scans it, and places it onto the trolley.

The scanner and its underlying software are pieces of logistical media, highlighting the importance of these media in organizing labor and increasing efficiency. Amazon is secretive about the software behind its operations. Because most used in the distribution centers is proprietary, patents filed by the company provide important clues concerning the nature of algorithmic management in the centers themselves. In 2004, Amazon filed a patent application for "Time-based warehouse movement maps." By tracking the pickers on their way through the FC, this software is designed to generate a temporal map of the space of the distribution center. The FC is mapped according to the time it takes workers to move between identifiable locations. The aggregation of travel times between points first generates a temporal map which then becomes the basis for algorithmic management of the distribution center, used for things such as "scheduling the picking of items, evaluating employee performance, organizing the storage of items in the warehouse, and other uses."[26] These relational maps constitute the distribution center as an algorithmically organized "timescape" governed by speed and efficiency.[27] Networked and wearable devices, like the handheld scanner dictating workers' pathways through Amazon's distribution center, clearly open up new possibilities for time studies designed to raise labor efficiency famously pioneered by Frederick Taylor, later complemented by the motion studies of Frank and Lillian Gilbreth.

Arguably, picking is the hardest job in the FC. Dictated by the rhythm of the scanner, some workers walk more than twenty kilometers in one shift. Workers in some distribution centers once had a countdown integrated into their scanners, stipulating the seconds in which they were obliged to reach the next pick. Tests of these countdowns in the close-by distribution center of Leipzig were disabled following a wave of complaints and protest, and so the Brieselang FC abstained from introducing them.[28] Nevertheless, like most workers in the distribution center, pickers are given clear performance targets varying between 60 and 180 picks an hour. Amazon argues that these targets are derived from earlier or average performances. To the workers, however, the criteria appear quite opaque and seem to increase over time: "Once you've reached your goals, they are almost always higher the next day or hour. I asked my supervisor once and he asked me to be sportsmanlike about it," reported a Leipzig FC picker.[29] Workers from Brieselang describe similar situations. Through the scanner, management knows every worker's performance down to the second. Leads and area managers regularly speak to workers with performance printouts in hand, asking them to maintain their pace or speed up.

Quotas, targets, and other systems of key performance indicators (KPIs) are crucial to not only the governance of mobility in logistics, but also the management of labor on the collective and individual levels. The distribution center is based on a system of real-time granular surveillance of every movement, while KPI constitute seemingly objective parameters by which labor can be measured and analyzed. However, how data collected by the handheld scanner and the FC's software system are used in Brieselang and elsewhere shows that quotas are both opaque and changing at the same time. In that sense, they become accelerating technologies, whereby it is precisely their often-illogical nature—once the quota is reached, it shifts, such that everyone is expected to perform better than the average—that allows for the micromanagement of labor via "feedback talks" and the increased production of relative surplus value. KPIs play a decisive role in the microeconomy of an FC, functioning as part of the seemingly neutral, abstracting and quantifying logic of algorithmic governance and standardized procedures—or, as one worker put it, "everything is standardized, the only thing that changes is the performance number"[30]—while simultaneously deriving their power precisely from their nonobjectivity. At Amazon, individual workers, teams, managers, and entire FCs are pitted against each other by way of performance indicators, trapped in a mode of constant competition. This impossible demand that every worker performs better than the average is an excellent example of the logic of constant acceleration implicated in the management of labor by seemingly objective quantification.

The scanner and its standardized procedures dictate paths through the coded space of the FC in an almost perfect example of the production of space characterized by algorithmic management and the mutual constitution of sociospatial practices and software. The scanner does not allow for any deviation in the work process whatsoever: "Sometimes the routes are really absurd and illogical, but you can't do anything about it as you have to follow the scanner. Sometimes I feel like it wants to keep me busy even if there is little work," reported another worker.[31] The picker position constitutes a prime example of a central quality of digital Taylorism: namely, the insertion of human labor into complex algorithmic machines in a way that inverts the relation between technology and human labor. Pickers serve as the executing body of the software, which dictates their path through the coded space of the distribution center. Jesse LeCavalier describes this process in his brilliant study of Walmart from an architectural perspective: "the collection of the workers constitutes an organic extension of the computer systems that control the environment but lack the dexterity and cost-effectiveness to execute the commands."[32]

The infrastructure of software and handheld scanner represents a peculiar step in this integration, which subjects human labor to the logic of software. A manager at Amazon's distribution center in Rugeley, UK, describes picking with the scanner as follows: "You're sort of like a robot, but in human form. It's human automation, if you like."[33] Referring to automation in his famous *Fragment on Machines*, Marx foretells an automatic system of machinery in which the worker is merely a "conscious linkage" between large systems of machines.[34]

In the case of the picker's job at Amazon's distribution center, one can place a question mark behind "conscious," as software organizes and controls the processes and the problem of automating picking is a very manual one: goods in Amazon's FCs come in very different shapes and sizes, which makes constructing a robot capable of grabbing all of them very difficult.

With this problem in mind, Amazon has pursued another route in its attempts to automatize picking—in 2012, it acquired the robotics company Kiva for $775 million. Renamed Amazon Robotics, the company produces small, sensor-equipped robotic vehicles which can drive autonomously and communicate with both each other and additional sensors installed in the ground. These robotic vehicles can lift shelves and drive them to where a product is needed. This innovation makes the pickers' long marches obsolete, and Amazon has equipped a growing number of its facilities with these vehicles over recent years. Here, the pickers are stationary and take the objects out of the mobile shelves that are driven to their position by the robotic vehicles. The specter of automation remains very present among workers in European FCs as well, and Amazon appears to enjoy issuing press releases about its efforts in automating its distribution centers during times of labor unrest and strikes.

After an ordered item is grabbed it from the shelf, the picker scans it and puts it into the trolley. Then, the scanner tells the picker the next shelf he or she needs to go to. Once the picking route is finished, the picker places the item onto another conveyor belt, which sends the goods through a sorting machine and on to the next station. The conveyor belts channel the goods to the lines with the packing stations situated next to the conveyor belts. Here, another set of workers takes the goods, scans them once more, and puts them into Amazon's typical brown cardboard packages. A computer screen shows the orders and the package to be taken from storage. The item or items are put into the package, paper or air cushions are added to protect the goods, and the package is sealed and put onto another conveyor belt. The package then reaches the "Slam" station where a machine scans it, weighs it—thereby checking if the correct articles are in the box—and automatically attaches a shipping label. This is the first time the customer's name appears in print. Another conveyor belt moves the packages to the shipping dock, where another worker loads them into waiting trucks.

The workers in "pack" also have quotas to fill, precisely measured through the scanning of every item in every order. As in the whole FC, the digital measurement system is completed by a tight regime of rather old-fashioned workplace surveillance. A small tower at the end of the packing lines is sometimes used by leads and area managers to check on workplace discipline. Talking to colleagues is discouraged throughout the FC, and even small breaches of routine are scrutinized. In the event of a violation, workers may be rebuked by a passing lead or asked to attend more formal "feedback talks" with their supervisors. In these feedback talks, which often follow even small breaches of protocol and dips in productivity, workers are sometimes issued so-called protocols detailing their faults. A protocol issued in 2014 in a German distribution center is not only an example of the meticulous ways in which workers are controlled but also illustrates the extent to which personal surveillance by supervisors plays an important role in Amazon's distribution centers:

> XY(employee) was on xx.xx.2014 in the time from 07:27 until 07:36 (9 min) inactive. This was observed by XA (area manager) and XB (area manager). XY was standing together with XZ (employee) between the receive sites 05–06 and 05–07 at the level 3 Conveyor in Hall 2 and they were talking. Already at xx.xx.2014 XY was inactive between 08:15 and 08:17 (2 min). This was observed by XC (lead) and XD (area manager). XY was returning from the toilet together with XW (employee) at 08:15. After that, she talked at workplace 01:01 in Hall 2 with XV. She returned to work at 08:17.[35]

Workers report being asked about the frequency of their toilet breaks and being cautioned about returning even minutes late. Breaks are a major topic of discontent in many FCs—walking to the canteen and designated break areas can take a long time in the huge centers and is often prolonged by security checks. In the worst-case scenario, this leaves workers with insufficient time to get food and return to their station on time. This has led to significant discontent among workers, whose protests pushed Amazon to add additional breakrooms at some locations.

STANDARDIZATION OF WORK/MULTIPLICATION OF LABOR

Labor in Amazon's distribution centers is standardized down to the smallest detail. Amazon has rules for every movement workers make, from the way they use the stairways to the correct lifting position, relying on time and motion studies in the tradition of Taylorism to develop standard operating procedures for the tasks in the FC. Here, personal forms of workplace control are accompanied by digital technology facilitating the precise measurement and control of labor in real time. Computerized optimization of labor processes produces a system of real-time surveillance organized according to various forms of benchmarks, standard operating procedures, targets, and instant feedback mechanisms. This standardization and decomposition facilitate not only a more efficient labor process but also a further flexibilization of the workforce.

This hints at a fact crucial to digital Taylorism as a whole: standardization of labor, the decomposition of tasks, and algorithmic management allow for the flexibilization and multiplication of labor. Training for most positions in the FCs takes a few days and is largely conducted by more experienced workers during running operations; in some distribution centers, it is a matter of mere hours. These short training times help flexibilize labor, which is crucial to managing the contingencies of supply chains in logistics. It is difficult to imagine how an FC workforce could double in the months before Christmas if the work processes involved required more skills or training. Nevertheless, this process is not without frictions, and the influx of thousands of new workers tends to be chaotic and stressful for more seasoned employees—but it works.

Short-term employment, seasonal work, part-time work, subcontracting, and other forms of flexible labor are still typical in the logistics sector—not only in the distribution center but at all points in the supply chain. Workers in the logistics sector, and especially seasonal and part-time workers, are often migrants working under even worse conditions than permanent staff. During the 2012 Christmas season, a report aired by the German television station ARD investigating the conditions of five thousand temporary workers, many hired by agencies from Spain and Poland, provoked a public scandal for Amazon. The Spanish workers profiled had been misled about pay and conditions, were not given contracts before departure, and were forced to accept worse conditions once they arrived in Germany. Some of the migrant workers were housed by an agency in ramshackle resort housing almost an hour from the distribution centers, to which they were shuttled in overcrowded buses. Late arrivals at the FC caused by these buses were deducted from workers' pay. The aggressive security guards at the resort, some of whom were neo-Nazis (as the report was able to show), proved particularly controversial and forced Amazon to end the contracts with the agency and the security firm.[36]

For Amazon, temporary contracts are clearly not only a tool to create a scalable workforce, but also a way to discipline and motivate temporary workers hoping to receive a permanent contract. The criteria determining which workers receive permanent contracts are complicated and opaque, but the hope motivates many temporary workers and thus generates higher rates of productivity. Outside of seasonal staff, a tendency can be observed in Amazon's German distribution centers to slowly increase the number of permanent employees over time, whereby the possibility of receiving a permanent contract also functions as a tool to govern and discipline labor. While temporary contracts and subcontracting are still important in Germany, these forms of labor are often even more ubiquitous in other countries.

A report on the 2014 Christmas season in Amazon's FC near Poznan by the Polish grassroots union Inicjatywa Pracownicza found that only six hundred of the three thousand workers were directly employed by Amazon and the rest were hired through temporary labor agencies (e.g., Manpower, Adecco, Randstad—all major global players in temp work). Most of these workers had very short contracts lasting between one and three months, and some were shuttled in by bus on trips of up to four hours.[37] In the UK, temp agencies like Adecco run their own offices in some of Amazon's distribution centers, managing their subcontracted workers themselves according to a "three/six-strikes-and-you-are-out" system. A temporary worker at Amazon's distribution center in Hemel Hempstead over the 2016 Christmas season explains the points system:

> 0.5 points for late arrival, or if you leave earlier or come late back from breaks; 1 point for any absence; even if you notify in advance or explain the reason it doesn't matter—you have the point anyway; 3 points for absence without prior notification; the accumulation of 6 points leads to termination of contract.[38]

So-called zero-hour contracts are another form of extreme flexibilization across UK warehouses. These contracts provide workers with no guaranteed working hours; they are called to work only when needed, often on very short notice via text message. Amazon hired twenty thousand temporary workers for their twelve UK FCs during the 2016 Christmas season. To get the workers, who often live quite far from the distribution centers to the FCs, Amazon and agencies operated shuttle buses at some locations. However, several workers described them as expensive and unreliable, and some workers even slept in tents in the woods close to the FC in Dunfermline.

In the US, Amazon has taken another step in facilitating the flexible mobilization of labor by starting the "CamperForce" program, seeking to attract workers living in their camper vans and traveling across the country. Amazon offers paid campsites and other benefits to workers willing to work in fulfillment centers during peak seasons. To keep the mobile workers from leaving this difficult job, Amazon sometimes pays one additional dollar for every hour worked if they stay until Christmas.

Reading job listings for the mobile CamperForce workers, one gets the impression that they address the hard nature of the work quite frankly. One can only speculate whether this is another measure to diminish the number of these particularly mobile workers leaving before the Christmas rush is over. For example, a 2017 Christmas season job listing for mobile workers at the FC in Campbellsville, Kentucky gives a good description of work at an Amazon FC. Workers "must be willing and able to work all shifts" and "willing and able to work overtime as required." Amazon describes the distribution center as a "very fast-paced environment" where workers must adhere to "strict safety, quality, and production standards" and "must be able to stand/walk for up to 10–12 hours."[39]

The mobilization and flexibilization of labor to handle the contingencies of supply chains is by no means unique to Amazon; they are widespread across the entire logistics sector, from Brieselang to Madrid and Long Beach to Shenzhen. The seasonal and mobile nature of many jobs demands various forms of mobility and flexibility from a large number of workers. The proliferation of a variety of contractual forms, outsourcing, and subcontracting is widespread across a broad variety of work in the logistics sector, characterized by a tendency toward hyperflexible and very short contracts that push contingency risks down the subcontracting line.

Furthermore, and this is quite clear at the Brieselang distribution center, flexible contracts are also a reaction to workers' struggles. At the Brieselang FC, temporary contracts are a major obstacle to joining other striking distribution centers. Many workers fear that even joining the union will diminish their chances at a permanent contract others have already accepted the idea that Amazon will be a relatively brief stopover in their work biographies. Many of the permanent employees, in contrast, feel privileged and are less interested in acting against their employer. Accordingly, the conditions facing militant workers and unionists at Brieselang are fairly difficult: while other distribution centers enter their sixth year of strikes, Brieselang is yet to join. But even if the balance of power at other FCs is more favorable for the union and militant workers, the workforce remains divided. After some initial public relations incidents like the ARD report on migrant workers being watched by neo-Nazi security guards, Amazon has improved its media strategy and proven to be a tough adversary for the union. In Germany and other places, Amazon attempts to obstruct union organizing and has responded by creating and supporting pro-employer "yellow unions" and lists for works council elections. An analysis of the strikes describes the antiunion activities by Amazon as "a lesson in counter-organizing."[40] As a result, the workforces at most distribution centers are fiercely divided between workers willing to strike and those aligned with their employer.

Nonetheless, the strikes have continued for many years since the first walkout at Bad Hersfeld in 2013, the first strike at an Amazon facility worldwide. Since then, workers in other European countries have followed. With the slogan "We are no robots," the workers are also protesting against the forms of digital Taylorism encountered in the distribution centers. Digital surveillance, pressure through KPIs, as well as the sometimes illogical and patronizing nature of algorithmic management have emerged as central issues of protest alongside questions of pay and collective bargaining, which were the central issues of the union in the beginning.[41]

Amazon often stresses that the strikes have little to no effect on the punctuality of their deliveries. Although there are many indications that the strikes do in fact have effects, Amazon's ability to reroute demand to other FCs to avoid obstructions has diminished their overall effectiveness. The new Polish FCs play a special role in this context. Against the backdrop of Amazon's ability to anticipate strikes and reroute orders, the union is now experimenting with short, spontaneous, and unannounced strikes at individual FCs instead of major national strike days. According to workers and shop stewards, these tactics have generated considerable problems and sent management into a panic on several occasions. Polish workers have also responded with a "go-slow-strike" in support of their German colleagues, hereby demonstrating that the struggle is increasingly becoming a European one.

The distribution center is a crucial node in the logistical supply chains operated by Amazon. From here, the packages start the last section of their journey: The delivery to individual customers. The process of delivery from the distribution center to the customer's doorstep has become another focal point of Amazon's operations, and the logistical reorganization of contemporary capitalism more broadly.

AMAZON'S NEXT FRONTIER: THE LAST MILE

In 2014, Amazon filed for another spectacular patent: A flying distribution center. The patent "Airborne Fulfillment Center utilizing unmanned aerial vehicles for item delivery" describes a distribution center designed as an airship.[42] It hovers at an altitude of approximately forty-five thousand feet and is designed to circle over populated areas to function as a base for autonomous drones, delivering to private customers in the area below. Smaller airships ("shuttles") are to be used to replenish the flying FC with inventory and transport workers to and from their airborne workplace. While further steps to implement this spectacular idea remain to be seen, the development of delivery drones is in full swing. On December 7, 2016, Amazon delivered its first commercial package via drone to a customer on the outskirts of Cambridge, UK. According to Amazon, the delivery by the autonomously operating drone took thirteen minutes from click to delivery. It was part of a private trial only open to two customers in the Cambridge area, where Amazon had been testing drone delivery since summer 2015. Besides high costs, legal aviation restrictions are currently among the biggest obstacles to automating the last mile through commercial drones in most countries. These restrictions, however, do not prevent Amazon and a range of other corporations like Walmart, DHL, Maersk, or Google from investing heavily in the development of such systems.

The reasons for doing so are not hard to comprehend. With the increasing importance of online commerce and the app-based ordering of almost everything, the requirements of capacity, speed, and flexibility on the so-called last mile of delivery have exponentially grown in significance. The last mile of deliveries to customers has become a site of extreme competition between many companies and the focal point of a far-reaching transformation that not only includes patterns of consumption but also profoundly impacts labor and the production of (urban) space. Showcasing its importance, an industry website describes the last mile as "the final frontier of logistics."[43] This is because the last mile is a highly complicated terrain, involving constantly changing routes and destinations, is cost- and labor-intensive, and increasingly important in the context of exploding demand for doorstep delivery.

Although Amazon has stayed clear of the complications of last-mile delivery for a long time, this has changed recently—one reason being that other delivery providers are simply unable to keep up with Amazon's requirements in terms of volume and speed. Amazon is continuously trying to increase the delivery speed of its products. In doing so, Amazon hopes to mitigate one of e-commerce's biggest disadvantages compared with brick-and-mortar stores: the time between the act of buying and receiving the goods. Its Amazon Prime subscription service has always promoted next-day delivery as a major selling point. In many areas, this has already changed to same-day delivery and, under certain circumstances, even same-hour delivery. For a long time, Amazon has used other providers to serve the last mile of delivery. In recent years, however, Amazon has started to push into the last mile of delivery by starting its own delivery operations. With this, Amazon has entered into a crucial section of logistical supply chains in contemporary times, and delivery vans sporting Amazon's logo have joined those of UPS, DHL, and many others congesting the streets of many urban areas. A glance at these city streets brings to light the ubiquity of logistical operations: they are swarmed by delivery vans of all sorts, bicycle messengers, food-delivery drivers on scooters, and many others trying to deliver all kinds of products to the customers at maximum speed.

The army of delivery vans and couriers on bikes and scooters is one very visible expression of a new logistical urbanism, whereby logistical operations move from the industrial parks on the city's outskirts into their centers. To offer fast delivery services, Amazon and other companies must move their distribution centers closer to their customers. Amazon has complemented its larger distribution centers, usually located on the outskirts of major cities, with smaller distribution centers located within city limits to serve as the starting point from which products can be delivered within hours. While logistical cities understood as logistics parks, ports, or special economic zones and their particular form of spatial and urban planning are most of the time situated at the margins of urban agglomerations, this imperative of speed tends to further merge the space of logistical operations with city centers.

In its foray into the last mile, Amazon is competing with a range of other businesses selling all kinds of products, from food to technical products. Same-hour delivery and app-based ordering recalibrate the city "as integrated service platform."[44] In this context, time becomes the most critical attribute of spatial production. Architect and urbanist Clare Lyster, who engages with how logistics reshape contemporary cities, argues that cities can no longer be understood primarily by static objects (as it is common for architects) but increasingly through its logistical systems and procedural flows, claiming that time is now "the most critical attribute of city making." "Logistics," she writes, "calibrates space according to time and thereby renders the city a timescape."[45] The idea of cities as timescapes resonates with the business of same-hour delivery and the labor of the drivers navigating the city in vans and on bicycles. Logistical flows of goods, information, and people are continuously reconfiguring contemporary cities. Hereby, the production of space is increasingly driven by algorithmic mobility systems—yet another type of logistical media—that are a crucial infrastructure of today's global cities.

LABOR ON THE LAST MILE

Despite all attempts at automation, the last mile remains one of the most labor-intensive sections of logistical operations. Labor in the last mile is situated at the intersection of a rising logistics industry and the so-called gig economy and therefore provides a crucial entry point into the analysis of the current transformation of production, circulation, and consumption. Situated at one of the most important and most expensive points of supply chains, labor in the delivery sector has always been subject to intense pressure and characterized by flexible and precarious labor arrangements. At the moment, however, it is subject to dynamic changes.

Two aspects are especially important. First, new forms of the organization and control of labor have been extended to delivery labor by means of digital technology. As in other areas of work, delivery labor is increasingly characterized by forms of algorithmic management, new technologies of standardizing and measuring labor, as well as intensified surveillance. Second, while the labor process has become increasingly standardized, the contractual and legal parts of labor relations has become subject to further flexibilization. The logistics sector—delivery, in particular—has always been characterized by outsourcing, subcontracting, and flexible labor contracts. With the emergence of the gig economy, however, this process has been amplified and intensified. As many important corporations of the gig economy (e.g., Uber, Deliveroo, Foodora) operate in the sector of delivery and transportation, platform labor has become an important tool in a sector that is already transforming labor relations in the industry. With its Flex program, Amazon has also started a gig economy platform for the delivery of its products. Labor in the last mile shares many similarities with distribution center labor—digital management, standardization, and tight surveillance on the one hand, and precarization and flexibilization on the other. Amazon's delivery drivers must compete with drivers from market leaders like DHL and UPS. The latter's truck driver labor illustrates that digital Taylorism is not restricted to the distribution centers but is ubiquitous across the logistics sector.

UPS is among the largest private-sector employers in the US, where it employs 374,000 of its over 450,000 workers globally. While UPS today is a differentiated logistics provider, with its own cargo airline and freight-based trucking operation, package delivery remains its core business. In 2017, UPS delivered an average of 20 million pieces a day, or a total of 5.1 billion packages, and generated revenues of over $65 billion.[46] Its iconic brown vans have become a major cultural symbol of the US economy, featured in various media formats. These vans are driven by more than 50,000 drivers in the US (and even more during the peak time before Christmas). A particularity of UPS (at least in the US) is the high number of directly employed drivers, which is due in no small part to the degree of union organization among UPS workers.[47] This has continuously limited strategies for maximizing profits by increasing flexibility in terms of labor contracts. Wages (and benefits) are also relatively high compared with industry standards, also due to the degree of unionization and long histories of struggles at UPS. Compared with corporations such as FedEx and Amazon, where unions have had little momentum, the nearly 280,000 workers organized at the Teamsters Union are an astronomical number. With a view to these particularities, it becomes clear that the intensification of work is of utmost importance to UPS to remain competitive. The sophisticated technologies employed by UPS to this end are another example of what can be described as digital Taylorism.

While relatively few full-time drivers at UPS have complaints about wages and benefits, long hours and the fast-paced, standardized, and disciplined nature of the work are a common source of discontent among drivers. UPS drivers have been working according to standard operating procedures for a long time. In training, future drivers learn a huge number of protocols on how to save time, such as how to start the truck with one hand while buckling the seatbelt with the other. The seventy-four-page guidebook handed out to drivers to maximize delivery efficiency is based on time and motion studies. These guidelines regulate the smallest details of drivers' labor, up to questions such as where to put their pen (in the left pocket for right-handed drivers).[48]

With the introduction of its "telematics" system, UPS has further radicalized the standardization and intensification of its drivers' work routines. Each delivery van is equipped with over two hundred sensors, and the drivers' handheld scanners (Delivery Information Acquisition Device—DIAD) produces additional data. The system collects a massive amount of data from the trucks (variables such as speed, braking, etc.), GPS data, customer delivery data, and driver behavior data. The system monitors things such as seat belt use, idle time, and how many times a driver backs up. Each time the driver stops, scans a package, or does any other thing, the systems record these details. A continuous flow of information is processed to UPS data centers, where it is collected, analyzed, and provided to supervisors.

The company knows precisely how much even small efficiency gains in their labor processes will benefit them: "Just one minute per driver per day over the course of a year adds up to $14.5 million," according to the company's senior director of process management Jack Levis, speaking to the National Public Radio network.[49] In public presentations, UPS stresses the savings in fuel and maintenance as a major benefit from telematics, but labor is clearly also a major issue. In a language both euphemistic and frank, UPS describes how the telematics system is used to manage labor:

> To maximize the benefit of telematics, we bring our drivers into the process. We give them and their managers detailed reports on how their behaviors stack up against the results we strive for, such as accelerating and braking smoothly to conserve fuel. Having concrete

> data empowers them to optimize their behavior behind the wheel and make their "rolling laboratory" even more efficient.[50]

The software establishes performance indicators, which are in turn used to apply pressure on drivers. "We have the driver data; we know how fast they're driving, how hard they're stopping," the director automotive engineering at UPS, Dave Spencer said more frankly in an interview with a business magazine. "That driver will change bad habits before it costs us money."[51] The strength of the union has helped reach an agreement that forbids UPS from firing workers based on low performance as evaluated by the telematics software, although UPS has found ways to work around this agreement, and many workers report how the metrics are used to pressure them. UPS drivers have reported that managers showed them printouts with details of their performance and asked them to increase their number of deliveries. Sensors installed inside the truck allow managers to scrutinize every break and even the style of driving; a printout of all the data generated by one driver during a shift can reach forty pages. Drivers are often forced to justify toilet breaks and even minor deviations from the rules to their managers.

Another important feature of the technologies of algorithmic management employed by UPS is its navigation and route planning system called "On-Road Integrated Optimization and Navigation" (ORION). The ORION software addresses a problem that appears straightforward at first but is in fact incredibly complex: finding the shortest route to connect a number of points in space. Even when the number of addresses is fairly low, the number of options rises very quickly. The formalization of the optimal solution to this problem, which came to be known as the Traveling Salesman Problem (TSP) in the nineteenth century, has become an important object of complexity theory, applied mathematics, algorithm theory, and computational geography. A brute-force computation of a route with more than a few stops is almost impossible.[52] ORION, however, stores more than 250 million address points, and a typical day tour of a UPS van includes more than one hundred stops. This is why even the ORION algorithm, whose code would cover roughly one thousand pages if printed, does not attempt to solve the TSP. Rather, it is a learning algorithm that works with automated feedback generated by the vans to provide a temporal map of its territory.[53] Such maps are key for an understanding of the city as timescape, exhibiting the importance of algorithmically driven logistics in the production of urban space.

Similar to the entire telematics system, ORION is focused on details and small efficiency gains, such as reducing left turns. However, efficiency for UPS is related to not only routes but also driver performance. An important issue for UPS is backing up. UPS prefers for its drivers to back up as little as possible, citing the increased risk of accidents. The telematics system monitors not only how often a driver backs up, but also the distance and speed at which it is done. If the software determines that a driver backs up too often, managers will ask that person to change their driving style. As one worker reports: "Our max backing speed is supposed to be 3mph. I got a message saying my backing speed was 3.7 mph on average and to please slow it down. I told them I would as soon as they installed a digital speedometer for me."[54] Like him, many drivers find the ORION software inefficient and patronizing, and many workers question the efficiency of algorithmic management compared with their predigital routines. Notwithstanding the question of which routine is more effective, such forms of algorithmic management take even the smallest decisions concerning how work is performed out of the workers' hands.

Software such as ORION is a tool to logistically map urban and rural space according to variables such as speed, distance, and fuel use; however, it also serves as a tool to increase pressure on labor and raise productivity with a multitude of targets and indicators. UPS workers report that with the introduction of ORION, targets have risen without the software managing to raise the efficiency of their routes, making it necessary to sprint or ignore safety concerns to reach the new targets. Just as in Amazon's distribution centers, the brown vans are nowadays part of a system of real-time granular surveillance of every movement, and KPIs constitute seemingly objective parameters by which labor can be measured and analyzed. In theory, KPIs play a decisive role in the micromanagement of labor, functioning as part of the neutral, abstracting, and quantifying logic of algorithmic governance and standardized procedures. In reality, however, quotas are often unrealistic and always shifting, and they thus become accelerating technologies rather than objective measurements of good performance.

The digital strategy to increase efficiency and further intensify labor works for UPS. Within the first four years after the rollout of the telematics system, the company was able to handle 1.4 million additional packages per day while the number of drivers had slightly declined.[55] The way digital technology allows for the measurement, organization, intensification, and surveillance of labor at UPS also shows how networked devices, sensors, and apps have moved Taylorist discipline as well as time and motion studies outside the enclosed spaces of factories and into the urban space of the logistical city.

RADICAL FLEXIBILITY: THE EMERGENCE OF PLATFORM LABOR

Unionized drivers employed full-time at UPS represent somewhat of an exception in the employment landscape of the last mile. Throughout the past few decades, however, UPS has been constantly trying to add more segments of part-time and fixed-term drivers. Despite union resistance, these attempts have been at least partly successful. Among the latest attempts in the direction of flexibilization is the company's idea to contract people using their own vehicle as delivery drivers, predominantly to answer increased workload in peak times. The idea itself is not original. Platform-based labor outsourced to independent contractors is becoming increasingly important in the last mile; several corporations started as typical gig economy platforms, and many older corporations have started to experiment with forms of hyperflexible platform-based employment. With its Flex program, Amazon copied the model with which Uber is disrupting the taxi market and introduced it into delivery. Rolled out in the United States in 2015, the program has continuously expanded and has also been introduced in countries such as Germany and the United Kingdom.

The term "Uberization" of delivery and logistics is misleading to a certain extent, as it suggests that such labor relations come into existence only through digital platforms. Rather, it seems necessary to reverse this narrative and to situate the logistics industry within the genealogy of the gig economy. In many ways, the logistics sector has always been a site of experimentation with hyperflexible forms of labor to find lean and cheap answers to the contingencies of global supply chains. Labor relations that are characteristic of the gig economy have been around in the logistics sector long before the advent of digital platforms. One example is the trucking sector in the US ports. In the late 1970s, the deregulation of the industry started and prompted the entrance of "owner-operators" or "independent contractors." These terms describe individual drivers who own or lease their trucks and contract their services to bigger freight firms.[56] Factually, these drivers are employees of these bigger corporations in almost all aspects except their legal status. Contracting drivers as owner-operators who are often paid by the piece allowed corporations to reduce wages and push many of the entrepreneurial risks onto the drivers, who are not entitled to things such as insurance, other benefits, or overtime pay. In 2014, approximately forty-nine thousand of the nation's seventy-five thousand port truck drivers were independent contractors.[57] These employment relations in the port trucking sector are in many respects an exact blueprint for labor relations we find in what today is described as the gig economy. It seems important to acknowledge such predecessors of today's gig economy, in order to gain a better, historically founded understanding of the continuities and transformations that characterize the current rise of platform labor.

"Be your own boss, set your own schedule, and have more time to pursue your goals and dreams. Join us and see how you can put the power of Amazon behind you," reads the advertisement with which Amazon aims to win individuals as "delivery partners" for its Flex program. The core of Amazon Flex is an app allowing people to register as courier drivers using their private vehicles. Following a background check, successful applicants can start working as independent contractors. The whole process is organized by the app, which must be installed on one's private smartphone and provides several instructive videos (instead of a training period). Once accepted, drivers can sign up on the app for shifts of one to five hours (so-called delivery blocks). Before the shift, it tells drivers where to go to pick up packages. At the distribution center, drivers get in line behind other cars, check in on the app, receive their packages, scan them, and start their delivery route organized by the app. Deliveries need to be confirmed on the app, sometimes including pictures of packages left at a doorstep. The app is not only a tool for navigation and scanning of packages; it is embedded into a software architecture that not only manages the labor process but is also designed to create a wider range of metrics (including customer feedback) to evaluate performance. These forms of algorithmic management of labor allow firms to substitute, for the most part, for direct managerial control over workers.

Workers, formally regarded as independent contractors, are promised earnings of at least $18–$25 per hour, and equivalent amounts in other currencies. While the pay seems good to many drivers, it quickly becomes clear that the $18 minimum is not the real wage. A Flex driver summarizes: "You think your making $18 an hour and tips but it all goes to gas and car maintenance. You put lots and lots of miles on your car."[58] Many also complain about the number of packages assigned for one shift. Should drivers fail to deliver them in the time designated, overtime is not typically remunerated; neither is the time it takes to drive from one's home to the various distribution centers and back home after the shift has ended. While the technology would be able to precisely account for these extra working times, Amazon is strategically forgoing these possibilities to save money. Furthermore, the drivers must cover their own insurance, taxes, and other costs such as social security. In general, real wages vary according to several factors but are usually far below the promised $18 and not seldom below minimum wages. Using the legal construct of the independent contractor hence helps Amazon lower wages while pushing extra costs such as for equipment or insurance as well as entrepreneurial risk onto the workers.

The first drivers began to sue Amazon in 2017, claiming they ought to be considered employees rather than independent contractors given that they are fully integrated into the business and the way Amazon organizes and controls their labor. The plaintiffs also argued that, after expenses, their earnings generally fall below the minimum wage.[59] Some of the lawyers representing plaintiffs against Amazon Flex are also involved in a class-action suit against Uber along similar lines.

Precarity is furthered by the fluctuating availability of work. Many Flex drivers complain about the insecurity. It seems that Amazon permits more drivers than needed, which often leads to bitter competition for shifts—highly typical of app-based algorithmic management of independent contractors, which is also a major problem for Uber drivers, Deliveroo riders, and other gig economy workers. In an online forum, one driver reported that "they [Amazon] continue to hire more and more people so competition has only increased. It has gotten to the point where the only way to acquire shifts is to obsessively be swiping one's offers screen all day."[60] Many drivers use auto-tap applications to try to get an advantage in securing themselves shift over workers using only their fingers, and pictures of smartphones hanging in trees in front of Amazon and Amazon-owned Whole Foods delivery centers have emerged: drivers hope to get more jobs by jumping milliseconds ahead of their fellow workers.

In case of complaints or problems, Amazon can dismiss the independent contractors far more easily than regular workers, a fact that is also a disciplining tool across the gig economy, where workers try to avoid complaints and go out of their way to keep customers and platforms happy, get high rankings and thus more work, and stay away from having their accounts closed, the gig economy equivalent of a dismissal letter. For Amazon, platform-based employment of independent contractors allows the creation of a highly flexible and scalable on-demand workforce with very low fixed costs. For the drivers, these employment arrangements are also flexible, which many value, especially those with additional jobs, but at the same time highly precarious in several aspects. Nonetheless, their number is growing. While it is hard to get exact numbers, there are clear indicators. The company-run closed Facebook group for Amazon Flex Drivers has already over twenty-seven thousand members. At the same time, a spokesperson for Amazon has given the sketchy number of "thousands of delivery partners" driving for Amazon Flex in the UK alone. To Amazon, these drivers are important to keep pace with customer demand and increase flexibility.

In many ways, Amazon Flex's employment model is not very different from employment relations that have existed in parcel delivery at the end of subcontracting chains for a long time. The digital platform, however, cuts out intermediaries and allows an intensification of flexibility. While it is clear how platform labor with its short "gigs" allows for flexible reaction to customer demand, an often-neglected aspect is how platform labor depends on the digital organization and surveillance of work to be effective and cheap. The various technologies of standardization and algorithmic management reduce training times and increase (automated) organization and control of the labor process, allowing flexible and short-term solutions in recruiting labor. It is precisely the new possibilities of algorithmic organization and digital control that make hyperflexible labor at the scale of Flex efficient, manageable, and scalable.

The rise of the last mile signifies important transformations in consumption patterns concerning things such as shopping and eating. These activities are crucial in how cities are built and navigated. Urban shopping areas and restaurants and the accompanying practices of consumption and everyday mobility (in other words, urban spaces) are subject to change by the rise of platforms. A very visible sign of this development is the rise of platform-based delivery of food. The urban landscape of Berlin, London, and many other cities is populated by an army of couriers on bikes or scooters working for food-delivery platforms like Deliveroo, UberEats, and Foodora. Many of these drivers are also independent contractors. Here, we encounter similar contractual arrangements and forms of app-based algorithmic management of the labor process. Some are paid by the hour, and others are paid based on their number of "drops" (deliveries), yet another form of reducing fixed cost for corporations and another tendency across the gig economy that might be described as the return of piece wages. Many of Berlin's bike couriers are migrants, often from crisis-ridden European countries, who can be integrated into delivery labor via apps easily even if they do not speak German, hinting at the importance of migrant labor across many logistical operations as well as how platforms are reconfiguring the stratification of the labor market. Despite these difficult conditions, the workers in food delivery have shown that resistance in the gig economy is possible. Recent years have seen a wave of struggles and strikes all across Europe, driven by inventive forms of organizing and striking, showcasing the challenge platform labor poses to unions as well as hints toward successful organizing platform workers.

The last mile is currently a focal point of such operations, where the rise of platforms meets the "logistification" of production, circulation, and consumption at this final frontier of logistics. In the context of a ubiquitous on-demand logic, the last mile has become an important factor in the time and the flow-driven remaking of urban geographies. This logistical urbanism is not only a matter of new transportation infrastructures, urban warehouses, or streets congested by delivery vans but also, for example, the future architecture of retail and public space in cities that is already changing by the rise of online retail and ever faster possibilities of doorstep delivery.

ANYTHING BUT SEAMLESS

"In the past, harbour residents were deluded by their senses into thinking that a global economy could be seen and heard and smelled," writes Alan Sekula in his *Fish Story*. "But the more regularized, literally containerized the movement of goods in harbors," he goes on, "that is, the more rationalized and automated, the more the harbour comes to resemble the stock market." To him, a crucial moment is the suppression of smell. Goods that used to smell are now stored in boxes "that have the proportions of slightly elongated bank notes."[61] The abstract character of containerized logistical landscapes is attractive not only to contemporary visual arts but to the logistics industry as well, where websites, brochures, promotional videos, and product catalogs are illustrated with a flood of containers, cranes, and dashboards. All of these pictures, videos, and narratives represent the logistical fantasy of continuous, seamless, and unobstructed circulation. Although these narratives can prove tempting when researching logistics and logistical media, the picture on the ground is quite different.

In August 2016, the Korean shipping company Hajin declared bankruptcy. Following the announcement, its fleet was stranded; some ships were held at harbors awaiting judiciary sentences, while others anchored in international water to avoid being confiscated. One day after another, crews were unable to find suppliers to sell them water and food. The electronics company Samsung, which had goods worth about $38 million aboard two stranded vessels, feared it would be unable to serve customers in the US during the critical months ahead of Black Friday. Hanjin's bankruptcy was a dramatic expression of the crisis of the shipping industry, which became visible after the 2007 global financial crisis. Ten years on, the global fleet of container ships had almost doubled in size, and freight prices are low due to massive overcapacity and harsh competition, threatening to bankrupt many shipping companies. As much as logistics is a means of shifting capital's contradictions around in space and time, it is also subject to and bearer of these contradictions. The container and the algorithm express the power of standardization and modularization. Bearing these images in mind, it is easy to understand this history as one of seamless and uninterrupted circulation and global homogenization. A central goal of this chapter is to upend this story by arguing for an understanding of logistics as anything but seamless.

A multitude of failures, obstacles, and systematic contradictions has fractured global logistics. On the ground, logistics is rife with obstacles, failures, problems, and misunderstandings. Many may first appear to occur for banal reasons, yet upon closer inspection, they can often be traced back to the contradictory and complex nature of the logistical task itself. These contradictions naturally include, but are not limited to, the central division between labor and capital, as another crucial contradiction emerges between cooperation and competition as internal to capitalist logistics. The result is a system of global circulation riven by differences and contradictions. This is also the understanding Sekula reaches in his visual and theoretical work on the forgotten spaces of container shipping.

As for the most important of these contradictions, that between capital and labor, recent years have proven turbulent. The logistics sector has been hit by a wave of workers' struggles, from the shutting down of the Port of Oakland—arguably the most powerful moment of the whole Occupy movement—to recent strikes and protests in Hong Kong and Valparaiso. In Europe, there are the conflicts of the Amazon workers in Germany, Italy, Poland, or France, and the powerful struggles of migrant workers in logistics hotspots in Northern Italy.[62] Workers throughout the gig economy are beginning to organize and strike; workers at Foxconn and other Chinese factories where many of the supply chains originate have fought for better conditions with some success for several years. There are many more sites where living labor has resisted its seamless insertion into supply chains. If one argument of this chapter is true—that the circulation of goods has become even more important today (i.e., if logistics is an ever more crucial industry to the global accumulation of capital)—then this is also good news for the workers on the docks, ships, warehouses, and trucks. Put simply, as the importance of their labor to capital increases, their (potential) power also rises dramatically.

3

THE FACTORY OF PLAY

Gaming

Steve Bannon, the former chief strategist of the forty-fifth US president, Donald Trump, once invested $60 million in a business based on the labor of digital migrants. In 2006, Bannon convinced his former employer, Goldman Sachs, to invest in a company called Internet Media Entertainment (IGE), which at the time was one of the most important actors in a shadow economy connected to massive online video games like World of Warcraft. This game was—and at this writing still is—one of the most popular of these games played online by millions of players from all over the world. Over seven million players populated the online spaces of World of Warcraft at the time of Bannon's investment. The game's digital world is a graphically impressive medieval landscape of dark forests, vast plains, green hills, towering mountains, wide seas, sprawling cities, and quiet villages populated by a multitude of human and magical creatures. Players kill monsters, explore the landscape, socialize, and complete quests, thereby developing their avatars' skills and accumulating gold and virtual goods to slowly advance through the levels of the game.

For those without the patience or time to do so, IGE offered gold, the in-game currency, in exchange for real money. The firm's website also sold virtual goods like weapons and rare "mounts" (rideable creatures), and even offered character leveling services—players could hand over their accounts and receive them back several hours later at any level they wished in exchange for money. Although forbidden by the game's publisher and frowned on by many players, the informal trading of real money for virtual goods is a multimillion-dollar business, and IGE was its biggest player in 2006 with offices in Los Angeles, Shanghai, and Hong Kong. Unfortunately for Bannon's investors, however, IGE soon ran into trouble. Players launched a class-action lawsuit against trading in-game currency for real money, claiming it would "substantially impair" and "diminish" their enjoyment of the game.[1] Beyond that, the game's publisher, Blizzard Entertainment, initiated harsh measures against the practice of real money trading, making it even more difficult for IGE to sustain its profit margins. Ultimately, IGE's virtual currency business was sold abroad, the investment failed, and the company reoriented and rebranded itself as Affinity Media, operating several gaming websites and communities. Bannon served as its CEO, a post he held until 2012 when he became chair of the infamous Breitbart News.

Like IGE, the entire shadow industry trading in virtual goods and currency took a hit at that time, but it continues to exist today in World of Warcraft and other games. After running into legal problems similar to IGE's, most Western platforms shifted operations to places where their supply was already coming from: Asia, and China in particular. By 2006, professional Chinese player-workers provided the overwhelming majority of IGE's inventory. These digital migrants to Western servers are commonly called "Chinese gold farmers." Effectively, Bannon had joined a company based almost entirely on Chinese (migrant) labor.

GAMING LABOR BETWEEN LOS ANGELES, BERLIN, AND SHENZHEN

At first glance, the gaming industry seems to be the workplace furthest from the Taylorist factory and enjoys an image as a liberal, creative, informal, and nonhierarchical sector associated with fun and play. This chapter, however, sheds light on the other side of gaming: the boring, repetitive, monotonous, disciplined labor of gaming such as the work of the Chinese farmers playing the game for hours on end to earn in-game items. Both the labor that goes into producing and maintaining games, as well as the labor spent in in-game economies, is to a large degree not creative and communicative but rather boring and repetitive. Of course, the gaming industry has its high-paid star designers and professional players who become famous e-sports celebrities playing in front of tens of thousands of fans, but they rest on the masses of testers and farmers who spend months in the same part of one game performing the same tasks over and over again. In line with the questions motivating this book, the labor of these people interests me most.

The protagonists of this chapter are, first, the so-called gold farmers in Chinese digital factories and, second, digital workers in game development and quality assurance in the sometimes huge corporations producing these games. Both groups of workers are part of the gaming industry's complex and transcontinental value chains. A focus on their daily grind produces a picture of the gaming industry as a digital factory rather than a fun playground for nerds refusing to grow up. The gold farmers, who are the subject of the following section, often work in locations that serve as a fascinating manifestation of the digital factory. Some of these gaming factories have hundreds of workers, mass dormitories, shift systems, quotas, and supervisors. Players alternate working shifts in front of computers where online games run twenty-four hours a day to accumulate digital goods that can be resold to mostly Western players who want to upgrade their experience.

While the first part of the chapter concentrates on the political economy of this in-game shadow business, the second part turns to the production of games. Developing a modern video game often costs several hundred million dollars and involves hundreds of workers across different parts of the development. This part of the chapter concentrates especially on a less glamorous part of this process: the job of workers in quality assurance and testing, whose job is most of the time to play the same sequence for days on end. They stand paradigmatically for a group of workers in the gaming sector whose labor is characterized by monotony and repetition as well as precarious conditions and long hours, labor conditions known very well to the Chinese gold-farming workers. Both testers and farmers are part of the multifaceted political economy of video games. Although profound differences are undeniable, striking similarities can also be observed, linking different forms of digital labor in the gaming industry.

"FOR 12 HOURS A DAY, 7 DAYS A WEEK, ME AND MY COLLEAGUES ARE KILLING MONSTERS"

After years of rumors and heated discussions in the gaming scene, it became clear in the early 2000s that gold farming had become a multimillion-dollar business employing a huge number of workers in factory-like settings. Initially, commercial gold farming was performed by individual players who discovered that money could be earned in their favorite games by accumulating digital goods and selling them to other players for real money. Some of these players began buying several computers and playing them simultaneously, while others formed groups to professionalize their living room- and basement-based businesses. Larger workshops began to appear in the early 2000s, most often located in China. According to rumors, South Korean entrepreneurs had begun exploiting China's low labor costs and hiring people to become professional gamers in popular online role-playing games like World of Warcraft. By 2008, studies estimated between four hundred thousand and one million digital workers were employed in gold-farming factories, the overwhelming majority located in China.[2]

The country emerged as a center of operations not only for its comparatively low labor costs but also for its huge gaming population—Chinese users constitute the largest national contingent of World of Warcraft players. Already well versed in the world of massive online video games, these players form a labor pool in need of very little training. Chinese digital factories were able to outcompete Western gold farmers easily, although a marginal number of farmers continue to operate in Western countries. In 2016, a Canadian farmer reported that he could barely survive with twelve-hour working days and that he made his biggest profits on major Chinese holidays when the gold price rises due to decreased output from Chinese farming operations.[3] Most customers are located in the United States and Europe, as well as Japan and South Korea—not to mention the emerging market in

China itself, as the new middle classes have more time and money to spare for gaming.

"For 12 hours a day, 7 days a week, me and my colleagues are killing monsters," a young Chinese worker described his job in the digital landscape of the Game World of Warcraft to a newspaper journalist in 2005.[4] The digital goods he can earn through this activity are sold by his employer to intermediaries owning digital platforms that offer these goods to bored players who want to advance swiftly through the levels of the game. His particular factory is situated in the basement of an old warehouse and shared with many other workers with similar tasks. A typical Chinese gaming workshop features twenty to one hundred computers and around fifty to two hundred workers playing in shifts so that every computer runs continuously. While some workshops consist of groups of friends trying to make a living out of their hobby, most operate in a professional and disciplined manner. Many of these digital factories offer dormitory housing and meals to their workers, who are almost exclusively men between the ages of sixteen and forty.[5]

Most Chinese gaming workers lack the necessary computers, game software, and accounts, as well as the language skills and PayPal account or other payment infrastructure to do business with intermediary platforms and Western customers; thus, they are forced to sell their labor to the owners of farming factories. Larger workshops, in particular, tend to operate in a highly organized manner; some employ supervisors, punch cards, and a shift system to keep the computers running twenty-four hours per day. The design of many of these digital factories can be described as a mixture between an internet café and a typical workshop in the same area. Some of the more professional workshops have uniforms, air conditioning, and motivational posters on the walls, whereas others are characterized by older computers, shabby construction, and difficult breathing conditions due to the heat generated by the computers. In smaller factories, the boss both directs the gaming workers and conducts business with clients, whereas bigger factories tend to have supervisors controlling the workers and exhibit a sophisticated division of labor. One gaming worker describes his digital factory to a gaming blog:

> The first gold farming company I was in was really big; I guess that this company owned at least 10,000 gold farming accounts. In my workshop there were 40 people who took turns to farm, some in the daytime, some at night. So the accounts are used for farming non-stop for 24 hours a day, 7 days a week. . . . Every day I feel very tired. You can imagine, every day I need to do at least 10 hours farming. I'm always looking at the computer screen and always seeing the same instance and the same mobs. So I feel very tired.[6]

Just as this worker describes it, tasks are often highly monotonous and exhausting, and shifts typically last ten to twelve hours. Although some workers are themselves committed gamers, the labor of gold farming itself is rarely pleasurable. While some farming activities involve more complex and exciting functions of the game, most farming is limited to very simple tasks. Like most other games, World of Warcraft has areas regarded as "farming areas" because the activities performed here allow users to accumulate gold or other items in relatively short periods of time. Farmers can often be found in these areas, as well. While observing them go about their business in the digital landscape of the game, one can clearly apprehend the repetitive and monotonous nature of farming. In most cases, a farmer will exploit one lucrative function or loophole in the game over and over. Many farmers, a surprising number of whom remain enthusiastic players of the game in their free time, report that their jobs are both monotonous and exhausting. One gaming worker argues: "You try going back and forth clicking the same thing for 12 hours a day, six or seven days a week, then you will see if it's a game or not."[7]

THE POLITICAL ECONOMY OF AZEROTH

World of Warcraft is probably the most well-known massively multiplayer online role-playing game globally. Since the first version was released in 2004, the game has profoundly changed the landscape of online video games. With peak subscription numbers of over ten million players monthly, many of whom spend the better part of their days immersed in the digital landscape of the game, it became part of the everyday lives of many players and quickly rose to a cultural phenomenon. World of Warcraft has expanded continuously since its 2004 release and has attracted more than one hundred million players over the past fifteen years. Today, the game runs on different regional servers all over the globe and is available in eleven languages. All players pay a subscription fee, making it a billion-dollar success for its publisher, Activision Blizzard. The California-based company, which has approximately 9,600 employees, encompasses several business units and publishes an even larger number of games for different platforms with diverse business models, including Call of Duty, Destiny, Skylanders, Diablo, as well as Candy Crush Saga and Farm Heroes Saga. World of Warcraft is among the most profitable of its titles, and the most famous frequently represented in other cultural formats (e.g., television series, books).

After players register for World of Warcraft, they must first design an avatar. Avatars are video game characters that typically develop over a longer time; they have a history, specific features, and a digital appearance. In the case of World of Warcraft, set in the fantasy world of Azeroth, various types of human and nonhuman avatars are possible. Most of them belong to either the Alliance or the Horde, the two great factions whose ongoing conflict structures the game. After entering Azeroth, the avatar begins its life as one of the millions of this magical medieval world's inhabitants. The player can now start killing monsters, exploring the landscape, and completing "quests." As in most role-playing games, the player's objective is to develop his or her avatar in various dimensions like strength, agility, spirit, and stamina. As World of Warcraft progresses, it grows increasingly difficult to move forward without the help of other players. Players thus link up to form guilds, the basic social unit found in the game. These guilds consist of groups of players, varying in size from less than ten to over one hundred fifty, who work together, speak via chat, and coordinate their activities to complete challenging quests.

The evaluation of an avatar is based on quite sophisticated metrics. Throughout its journey through the various levels of the game, the character must collect strength and experience, weaponry, a mount (rideable creatures, very important in the game), and much more to progress. The key to obtaining items such as weapons and armor is the in-game currency, gold. Gold can be looted from various creatures and other players, but it can also be earned by certain activities such as picking herbs, mining metals, or fishing. The fruits of the avatars' digital craft labor can be traded at auction houses. Enjoying the game is often determined by one's digital possessions, but these possessions must be earned, often by fulfilling tasks that are not particularly pleasurable. While much of the play involves complex and often cooperative and exciting actions, dull, repetitive, and monotonous activities also constitute a huge component of World of Warcraft. Particularly in the lower levels, the task of advancing one's avatar is characterized by a playing experience that is often "mindless and repetitive to the extent that it verges on Taylorism. There is an assembly-line mentality to many of the quests," as Scott Rettberg, a scholar in digital cultures, observes.[8] There are few ways of avoiding these strenuous parts of the game in accumulating gold and possessions and advancing through the levels of World of Warcraft. Gamers consider this system fair, as the absence of shortcuts allows the in-game possessions to represent the time and skills players have invested. For players who grow frustrated with the "grind," and want to catch up with their friends, or who long for rich and powerful avatars, however, this system is a problem. This problem, in turn, establishes the basis for professional farmers.

A DIGITAL SHADOW ECONOMY

Leaving aside a late-introduced and limited feature, World of Warcraft does not officially permit the exchange of its in-game currency and items for real money to provide a level playing field to all users. This prepares the ground for the gold-farming shadow economy in World of Warcraft and other games with similarly closed virtual economies. There are various forms of gold farming. The classical form is to play the game to earn in-game currency—gold, in World of Warcraft's case. A gold farmer logs into the game and performs certain tasks with his or her avatar. A typical task in World of Warcraft would be to kill a group of enemies, who then drop gold and other valuable items. The farmer takes these items and hoards them on the avatar's account. Through various platforms and intermediaries, this gold is later offered to players willing to invest money to buy digital gold. Another service offered is what many gamers call "power leveling." Here, game workers develop an avatar through various levels of the game, accruing power, skills, weapons, and gold. Afterward, the account is handed over to the buyer, who saves time and effort developing his or her character through the lower levels. A third and more sophisticated form of farming is to assemble up to sixty game workers to form a group of high-level players. The group of professional gamers is then sold as an army of virtual mercenaries to customers who need support fighting the elite monsters in World of Warcraft's final levels.

A simple Google search reveals a wide range of sites selling World of Warcraft gold, currencies of many other games, as well as digital items or even developed avatars and other products. One of these shops is the German market leader MMOGA.de. The site offers a wide assortment of gaming services and claims over seven million customers. For World of Warcraft, it offers gold on all servers with a variety of possibilities to get it handed over as well as power leveling. "Your character is only levelled by professional gamers in order to reach the desired goal quickly," promises the platform. "We intermediate only the best professional players to you," the advertisement goes on and claims that these professionals will "not use third party programs or bots which could threaten your account."[9] MMOGA was sold to a Chinese corporation for €300 million in 2016.[10] This sum—and the fact that practically no one took notice of the deal—shows that the industry is huge in terms of revenues while remaining largely in the shadows. This is due to the legal gray areas in which these platforms maneuver. While selling gold for money is clearly not allowed by the publishers of World of Warcraft and other games, the precise legal situation is less clear, as laws in most countries are poorly equipped to deal with the sale of virtual items and currencies for real money.

Furthermore, platforms typically act as intermediaries between farmers and customers. "We're like a stock exchange. You can buy and sell with us," as Alan Qiu, founder of the Shanghai-based gaming workshop Ucdao.com describes it. "We farm out the different jobs. Some people say I want to get from level 1 to 60, so we find someone to do that."[11] This is an attractive option for many players with little free time. In World of Warcraft, it takes an average person several hundreds of hours of play to reach the highest level. "We purchase the gold from tens of thousands of farmers. And we resell it via retail platforms. So to some extent we are an exporter," as an employee of another large platform for the sale of virtual goods and currency breaks down the business model. "The only difference is that the goods are virtual and the procedures are operated in a digital environment."[12] The owners of the gold farms, and even more so the intermediary platforms, end up securing the biggest profits.

Because it is a global shadow industry, it is difficult to estimate revenues in the gold-farming sector. During the heyday of Steven Bannon's old firm, IGE, sales estimates in virtual goods ranged from $300 million to $10 billion.[13] The industry operated more publicly at the time, and IGE hoped to reach agreements with the publishers of the biggest games. The agreements never materialized, however, and publishers soon began taking stricter measures against gold farming. IGE moved its business to Hong Kong, and its US branch was barred from selling World of Warcraft virtual goods by a settlement reached with disgruntled players. The Hong Kong branch managed to stabilize and was among the most popular online stores for digital gold and other items for several years before it sold off these operations and went offline for unknown reasons. These turbulent years solidified the status of gold farming as a shadow economy and made it considerably harder to gather reliable data and information on the industry's overall size and development. Some platforms attempted to concentrate on their intermediary role, thereby pushing much of the risk onto the sites where farming labor is actually performed: the digital factories of game labor.

DOUBLE MIGRANTS

The composition of the Chinese farming workforce has changed over the years, as some first-generation workers, mostly students who did their farming in internet cafés, opened their gold-farming workshops and increasingly employed migrants from rural areas. Shenzhen farm owner Wei Xiaoliang is quoted in the *South China Morning Post* as explaining they "prefer to hire young migrant workers rather than college students. The pay is not good for students, but it is quite attractive to the young migrants from the countryside."[14] Some of these gold farmers were once actual farmers, before becoming migrants to China's booming cities. Within the online spaces of World of Warcraft, they become migrant workers for a second time.

A 2011 documentary, Ge Jin's self-made and fragmentary *Goldfarmers* provides rare insight into the lives of these Chinese digital laborers.[15] While Ge Jin's low-budget documentary is particularly interesting for its insights into the working conditions in several Chinese gaming workshops, it also concentrates on another aspect of the labor of gold farming. Its special focus sheds light on the experience of the farmers in the game and their interaction with Western players—and these encounters are often hostile. A Chinese game worker featured in *Goldfarmers* reported, "If they know you are a Chinese farmer, they would say you have no right to be here, or even attack you without reason."[16]

Although many Western players use the services of gold farmers to advance through the game, World of Warcraft's general culture disapproves of commercial farming and the selling and buying of gold for real money. It is considered cheating and goes against the ethos of the game as well as its in-game economy, as players perceive that it causes inflation. Hence, gold farmers are regularly attacked for their activity. The digital workers doing gold farming know the hatred they evoke in many Western players very well. Sometimes being passionate players themselves, they know how their work can interfere with the play of leisure gamers. One worker featured in *Goldfarmers* explains:

> A professional gamer normally stays in one spot and kills the same monster, over and over again, so that he can keep getting gold. Because this is his job and there is pressure from the boss, he has to stay there. If some other players come to the spot he has no choice but to fight with them. Because he has to work and he is under pressure. So we professional gamers do have an impact on regular gamers. . . . If you see a professional gamer in the game, I wish you can understand his job and give him a little space. He will be very grateful. He will not go to other spaces and disturb you. He only needs a little space.[17]

Farming has become profoundly racialized within the space of World of Warcraft. Not all farmers are from China, nor are most Chinese

World of Warcraft players gold farmers. However, in the game's terminology the expression "being Chinese" or "playing Chinese" has become synonymous with gold farming. As the physical body cannot serve as a point of departure for the process of racialization within an online game, the main marker becomes a specific style of playing, or rather laboring in the game. In many cases, farmers are fairly easy to spot, often remaining in the same lucrative location and performing the same repetitive tasks to earn gold. Avatars behaving in a way that suggests they are not playing but working, or even just remaining in spots known for their farming possibilities, are potentially subject to racist attacks. Farming labor has thus become profoundly racialized, and the backlash against farming practices is fueled by racist imagery and slogans.

Some workers' only task is to hand over the gold to the buyer's avatar or to attract new customers in the game, all without being detected by vigilant players or the "Game Masters," Blizzard's in-game police. Accordingly, gold farmers' behavior is often likened to that of offline street drug dealers—a figure which almost always appears as a migrant in Western imaginary. Advertising for cheap gold in the chat channels is often the source of resentment against farmers; these ads prompted Blizzard to introduce an antispam system that makes selling via in-game messages quite difficult. In World of Warcraft's digital space, gold farmers are addressed as illegal migrants working in a space where others play. Throughout the game's landscapes, a constant form of racial profiling occurs to differentiate between legitimate "leisure players" and unwanted "player-workers." Avatars whose names are composed of numbers or appear to not be "Western" in any way are oftentimes treated with suspicion, as are players who do not react when spoken to. Western players even form "vigilante" groups to hunt down "Chinese farmers."

The documentary *Goldfarmers* also profiles Western players and their reaction to gold farming. One of these players is Gareth, a hardcore player in the US who runs a radio show dedicated to the game. He also runs a website opposing what he calls the "gold selling industry." Asked about his opinion on gold farmers, he argues they "don't have respect for the intellectual property rights of these companies and they also don't have respect for the other players of the game."[18]

He describes various groups of Western players operating websites with lists of well-known farmers and their typical locations, "so that people can actually go out when they are bored and kill these farmers and stop them from collecting money and items."[19] "Killing" means killing or obstructing the avatar used for farming, making it harder for workers to do their job.

Players have uploaded some videos of themselves attacking farmers onto YouTube with titles like "Chinese Gold Farmers Must Die," and others have even produced songs and videos against Chinese gold farming. In an excellent paper, Lisa Nakamura shows how these videos contribute to the racialization of labor in World of Warcraft.[20] Gareth, the player from *Goldfarmers*, describes a raid against farming workers facilitated by his radio show:

> I did a show on World of Warcraft Radio where we went to a popular place called Tyr's Hand in Eastern Plaguelands [a spot in World of Warcraft], a very popular place for farmers because of the drops [the gold and items that can be looted from creatures once they are killed] there. During the show, we had about 20 people go through, camp the area, and basically kill farmers every time they would respond. That was a lot of fun.[21]

DIGITAL LABOR/DIGITAL MIGRATION

These Chinese workers inhabit a strange dual position. While they stay in their country of origin and belong to the emerging digital working class of the Global South, within the space of the game and its surrounding culture they make many of the experiences and exhibit almost all qualities of "real-world" migrants. They enter the space of the game as workers in a position different from the hegemonic culture of the space itself. Their labor is sold to Western players as a service, which is also the reason for the attacks from other Western players. To work where others spend their leisure time is a common trait of migrant labor in a variety of "offline" professions, often in the service sector. Both in their representation and economic position, gold farmers inhabit the position of the migrant as well as "cheap labor" from the Global South. Interestingly, these dual positions are manifested in one person: a gold farmer is both outsourced labor power in the periphery, as he or she works in a Chinese workshop, and also a migrant doing the dirty work of farming while being subject to racist attacks within the digital sphere.

In an important ethnography on the mobility of Indian IT workers, A. Aneesh coined the notion of *virtual migration* to shed light on the new forms of labor mobility enabled by digital technology.[22] While many Indian digital workers in the IT sector migrate to Europe, the US, or Australia, others remain in India but work for Western companies. A variety of flexible and temporary models are aimed to match the factors of labor and infrastructure costs with the demands of being on-site with customers. These processes connect Indian IT hot spots (and their specific patterns of local and national mobility) with other global sites in multiple and complex ways, adding a further dimension to common outsourcing practices. Aneesh uses the term *virtual migration* to describe the experiences of the workers staying in India while working for customers abroad, whose labor is situated in cultural, spatial, and temporal contexts that do not match their physical location. He argues that these Indian digital workers who work for Western companies without leaving India "migrate without migration."[23]

The notion of virtual migration is important as it shows that in the context of networked economies, geographies are set in motion in such a way that concepts like outsourcing fall short of the spatial, social, and economic complexities at play. Maybe even more than the case of Indian IT workers, the political economy of online games and the experiences of the Chinese gaming workers shows the familiar concepts of outsourcing and offshoring fall short of the realities and experiences of networked digital labor. The online multiplayer game is both a global economy and a shared lifeworld, in which many leisure and professional players spend the majority of their waking hours. The farmer's plea to Western players to understand the gold farmer's job and grant him a little space in the game shows that the interactions in the game have affective as well as economic implications. For individual workers, attacks make the job harder—both emotionally and in terms of the daily quotas most workers must fulfill. The material risks, vulnerability, and the affective dimension of working in illegal or informal economies are characteristics digital migrants share with many of their offline counterparts.

The shadow economy of digital gold farming, and the digital migrants it produces, represent a new topological geography brought forth by new forms of global infrastructural connections. The nature of digital labor, or the nature of its product, exhibits a certain quality that complicates the demarcation between the categories of the mobility of labor and mobility of goods. Networked infrastructures and software allow the global transmission of data in milliseconds so that, for example, different laborers can work on the same project simultaneously from two different continents. This produces complex spatial formations that challenge common terms like *outsourcing* and *offshoring*.

Here, a theoretical vocabulary of virtual or digital migration may prove helpful in thinking further. Using the vocabulary of virtual or digital migration, I do not argue for a radical break with forms of outsourcing or offshoring. Rather, these concepts can be added to existing vocabularies to better understand the current transformations of labor mobility enabled by digital technologies and infrastructures. We can see how digital technologies and infrastructures challenge not only notions of topographical economic space but the related questions of labor mobility and the multiplication of labor as well. They are part of an ongoing heterogenization of global space constituting fragmented, overlapping, and unstable cartographies and questioning stable categories such as North/South or center/periphery.

The rural migrant working in a digital gaming factory on the edges of Shenzhen and in the digital economy of World of Warcraft is not only a "double migrant," but also inhabits a complex economic topology. The sites of labor power, of labor, of consumption, and the site of the buyer operate at various disparate but overlapping levels, tied together by the political economy of World of Warcraft, internet infrastructures, various forms of brokering through intermediary platforms, payment systems, and so on. These complex and fragmented spatiotechnological formations correlate with multiple and fragmented figures of labor and, thus, with multiple and fragmented figures of migration. The gold farmer is perhaps the paradigmatic example of this reality.

AFTER THE GOLD RUSH

The heyday of Chinese gold farming in World of Warcraft may be over. Although a great number of platforms continue to offer World of Warcraft gold, the business has diversified. Leisure players and the game publisher's attacks on gold farmers have had economic effects on these workshops, some of which were forced to closed through bans on accounts and IP addresses. As in World of Warcraft, gold farming is forbidden in many games, and gaming companies try to fight it as much as possible. Blizzard, World of Warcraft's publisher, employs a large number of technical experts who try to safeguard the game against farmers and bots, while many "Game Masters"—a sort of in-game customer service employed by Blizzard—spend most of their time hunting farmers and closing down suspected farming accounts. Blizzard bans thousands of accounts suspected of gold farming every month, although this line of action is directed almost exclusively against the farmers and not their customers. Consequently, actions against farming have repeatedly forced Chinese workshops to shut down, lay off their workers, and sell their computers.

According to rumors in the gaming community, a wave of hacked accounts flooded the market for a time, until Blizzard took measurers to secure their gamers' accounts—arguably making World of Warcraft accounts safer than bank accounts, at least for a while. Besides account bans and other repressive measures, Blizzard also charted a new course against farming by partially legalizing this business model. In 2015, Blizzard introduced the World of Warcraft Token, which can be bought from Blizzard and then be sold to other players in the auction houses for gold. A player who buys the token at an auction cannot convert it back into hard currencies but receives a month of free play. By introducing this legal way to buy gold, Blizzard takes a share of the trade in both gold and currency, while simultaneously undercutting the gold-farming business model. The token is of little use to them, as it does not permit converting World of Warcraft gold into cash. The token leaves some room for the gold-farming industry to maneuver, particularly if it manages to outprice the token, but this room is growing increasingly tight in World of Warcraft.

Rising wages in South China also work against digital factories in these areas. The area around Shenzhen in the Pearl River Delta has witnessed a wave of workers' struggles demanding a bigger share of overall profits, and the region is no longer able to compete on the lowest wage strata. Gold farming continues to be an issue around the gaming world. In recent years, for example, players in crisis-shaken Venezuela have turned to various games such as Runescape to generate some cash. In the context of hyperinflation, gold farming provides a way to generate money in a foreign currency to make ends meet. This farming has created tensions in Runescape, and guides on how to kill Venezuelan farmers in the game have appeared on the online forum Reddit, starting a controversial debate among gamers.[24] Many criticized attacking Venezuelan farmers who were just trying to survive financially, while others insisted on the illegality of their activity. Many of the Venezuelan gaming workers, however, feel that they have few other options, as one impromptu gaming worker reported to an Australian gaming website: "I have friends who play daily and if they do not play, they do not eat that day."[25]

The experience of these workers in China, Venezuela, and other countries as virtual migrants to foreign servers most clearly demonstrates the complex topologies of the mobility of labor in digital capitalism. Online games become economic spaces, thus also becoming sites of labor and exploitation with complex and fascinating politics revolving around in-game currencies, farming labor, and conflict. The spatiality of these games activates complex value chains based on global inequality and enabled by various infrastructures, including not only servers and fiber-optic cables but also intermediary platforms and global payment infrastructures. Furthermore, the shared lifeworld populated by both leisure and professional gamers sets in motion another set of politics around the racialization of informal labor. It is clear that digitalization is profoundly reconfiguring labor and mobility, and that digital migration will continue to become an even more important form of labor mobility.

These forms of digital migration must be viewed in continuity with other forms of circulation. Goods increasingly travel not only on ships and planes but also through transcontinental fiber-optic cables, which has profoundly reconfigured the global geography of production and circulation. Today, the journey of labor power takes many forms, most obviously that of hundreds of millions of migrants crossing borders or moving to urban areas in search of better lives. In addition, global logistics and infrastructures allow labor power to travel the globe crystallized in commodities; communication systems allow even faster transmission of data and services over great distances. Compared with these very visible forms of the mobility of people and goods, the forms of virtual migration undertaken by farming workers in online games and explored in this section are less evident. A Venezuelan worker who is racially abused and attacked as someone "who has no right to be here" while he works in an informal sector providing a service to the very strata of players that attack him shares more with many of the migrants who have crossed a territorial border than is obvious at first sight.

PRODUCING GAMES: LABOR AND CONFLICT IN GAMING STUDIOS

Now we change the scenery. Our new location are the offices of a gaming corporation in Berlin. Here, a game very similar to, if considerably smaller than, World of Warcraft is produced and maintained. The offices belong to a major German gaming company that was acquired by a Chinese gaming firm in 2016 but continues to run as its independent subsidiary. The company opened its Berlin offices in 2010 by acquiring another gaming studio that went bankrupt. Around sixty employees occupy the sixth floor of an office building close to Berlin's famous Alexanderplatz. The rooms are not very different from other office buildings nearby, even though they are equipped with some of the typical features of the gaming sector such as big screens, consoles, beanbag chairs, table tennis, and energy drinks. Despite these typical new economy artifacts and a decidedly casual atmosphere, these offices are also the site of a labor conflict.

One can find literally only one piece of paper in Smalline's Berlin offices—it belongs to the works council and is stuck to its blackboard in the hallway.[26] "German industrial law stipulates that any information regarding the works council must be posted in paper," explains the works council spokesperson, "but all our internal communication is online, on the intranet."[27] In his opinion, this is paradigmatic of the relation between the laws and regulations concerning workers' rights and the gaming sector: "Unions and strikes are considered to be incredibly outdated by my colleagues. They use paper and personal meetings. No emails, no videoconferences, this is considered to be old-fashioned and antiquated."[28] And yet the works council he leads is a unique success story, the first of its kind in Europe. In an industry where strikes and unions are practically unheard of, the existence of a works council, standard in many "old industries" across Germany, remains quite special.

The opportunity to open the works council at one of the bigger German gaming corporations arose when the gaming bubble burst in 2012—an event that went relatively unnoticed by the public but spelled a crisis for almost all European game producers. Smalline laid off almost one hundred employees, sending the others into panic. The person who is today the works council spokesperson was also on the list to leave the firm. His temporary contract had expired, and the company did not intend to keep him. However, he found a loophole in German labor law that requires a company to send a worker home after his or her temporary contract runs out or else the worker becomes permanent. The turmoil of the crisis worked to his advantage: "They had also fired many of the HR people, so it was all very chaotic," he explains. "They didn't know that law, and didn't send me home, so they had to give me a permanent position. One hour later I started the works council."[29]

The success of his initiative remains rather singular in the German gaming industry. When workers at Smalline's competitor Supgame Studios began planning a works council, their employer reacted with intimidation and pressure.[30] Two different groups of workers had contacted the union ver.di to discuss the possibilities of a works council in 2015. In autumn of the same year, the union secretary visited Supgame's offices in Hamburg to gather an impression of the situation there. Her visit did not go as planned. When she arrived that day, the workers who invited her met her in the streets before the offices: "They had been told that they were fired this very morning, they had to pack their stuff and were escorted off the premises immediately. They were completely surprised and devastated."[31]

Supgame claimed the terminations were unrelated to the works council initiative, but both the union and the workers knew better.

Many of the sacked workers had begun familiarizing themselves with procedures and were discussing the best way to launch the works council on an office messenger channel. As it turned out, many of the twenty-eight dismissed employees had also been members of this chat. Supgame heaps praise on itself for its campus equipped with a swimming pool, free food, and parties, but many workers complain about the lack of holiday time, tight schedules, and low pay. When Germany established a minimum wage in 2015, this in fact meant a pay raise for many at Supgame. Even developers with university degrees had been working full-time for less than €2,000 a month. After the twenty-eight workers were fired, discussions in the workplace continued and many employees reported intimidation and attacks against the union as a "foreign element that wanted to harm the company."[32] Finally, a meeting was held in early 2016 to establish a works council, but more than half of the employees ended up voting against the process necessary to move forward. The council might have proven useful six months later, when Supgame terminated around five hundred workers—roughly half of its total workforce.

TESTING LABOR

Smalline's main offices are also located in Hamburg, Germany's unofficial gaming capital, with other offices in Malta, Lyon, Istanbul, Seoul, San Francisco, and Berlin. Compared to Supgame's campus in Hamburg, the Berlin branch of Smalline is significantly less extravagant. The Berlin office's sole task is to maintain and develop the game Dragonvoice Online.[33] The game is an online multiplayer like World of Warcraft but differs in its business model. Unlike the subscription-based World of Warcraft, Dragonvoice is a free-to-play game. Players access the game in their browser, choose an avatar, and begin playing. Dropped into a medieval and magical world, the avatar fulfills quests alone or together with other players to gain strength and advance through the levels. Even more so than in World of Warcraft, progress depends on the amount of in-game currency players can accumulate. This is the key to Smalline's business model. As there are neither subscription fees nor advertisements in the game, the sale of in-game currency is the publisher's sole source of income, which requires the game's development teams to carefully manage levels of player frustration. The game now has more than seventeen million registered accounts and more than two million users who play the game on a regular basis, with sixty to seventy thousand playing at any given time. Only a small percentage of these players buy in-game currency. These players, however, spend enough to make the game a profitable operation.

Throughout my research, one team in Smalline's division of labor attracted my attention: the gaming workers in the quality assurance (QA) department. The QA team's task is to search for errors in the game and report them to the software engineers. A new version of the game is completed and uploaded to the test servers approximately every two weeks, which QA workers must then play through to look for errors. "You test the exits, the entrances, the various rooms, the virtual drinks, weapons, movements, everything," explains a QA worker.[34] If the testers find an error, they complete an "error ticket" and send it to the responsible team to fix it. Testing labor tends to be highly monotonous and repetitive. "It is like actually playing the game, only that you do the same stuff all the time. That's very tiring."[35] Even if some QA employees enjoy playing the game in their free time (surprisingly common among workers), they find testing exhausting. "I worked in a steel factory for a year and it was just as exhausting as working here. After you have spent the whole day trying to reproduce the same error and you have clicked your computer mouse at least seventy thousand times, you are quite tired."[36]

Usually the first group to arrive in the morning, the QA workers test the batches completed the day before. Dragonvoice was developed from scratch in Berlin; the game comprises over half a million lines of code and is constantly being reworked and enlarged. Beyond coding, testing is also extremely labor-intensive, as it is very difficult to automate and must be performed by human workers. At the same time, most of this labor does not require extensive training or extraordinary creativity and is thus highly interchangeable. Most QA workers lack formal training, although a few have received a certificate from a six-week training course. Generally, the team features two types of employees. The first type is the core team, most of whom have been working at Smalline for some time. Many of these workers remain in the job for years, and some hope to advance to other jobs in the company. Alongside them is the second type: the short-term employees needed for their fresh view of the game. They are referred to as "testing monkeys." Those workers stay for a few months and are then replaced by a fresh group. Many are interns, all of them are fans, and they often aspire to work in the gaming sector themselves. "Some are surprised that there is any pay. They are very happy to come here and see how it works. At least in the beginning,"[37] explains the head of the works council, who also works in QA.

This is not only the case for the interns at Smalline's QA team but throughout the whole sector. "Ninety percent of the employees are fans of the gaming sector," continues the works council spokesperson at Smalline's Berlin offices overlooking Alexanderplatz, concluding sarcastically: "The zealousness to be part of this great project has sectarian traits."[38] This increases many workers' willingness to accept low pay and long hours and fosters a mentality that makes it hard for unions to gain ground in the sector. The union secretary responsible for Supgame Hamburg corroborates this impression: "They are all fans, they really want to work there. For this reason, they are often ready to accept low wages and bad working conditions."[39]

In Smalline's case, the beginning of worker self-organization was fueled by layoffs, although pay also played a role. The initiator of the works council explained to me that "the main reason why I founded the works council is that I earn €5 an hour."[40] The €5 wage was well below any of the proposed minimum wages being discussed in German politics at the time, and could only be found in a few other sectors like construction or cleaning. Typically, QA workers constitute the lower strata of labor in gaming corporations. Workers are often motivated by hopes of advancing to other jobs in the firm, which makes them more inclined to accept low pay, long hours, fixed-term contracts, and often very exhausting and monotonous forms of labor. In the gaming sector in the United States, the situation is comparable: the testing workforce is composed of mostly young workers, most of whom have temporary contracts and feel that they belong to the lowest and most expendable class of workers in gaming corporations.[41] In a contribution for the magazine *Jacobin*, a former QA worker at Red Storm, a prestigious North American studio in North Carolina, reports that many young workers joined the QA department at Red Storm hoping to work their way up the ladder in "the new Hollywood of the video game industry." Most start with temporary contracts and pay at minimum wage and expected workweeks of sixty hours in peak times. The fluctuation is high, and often temporary workers are let go at the end of a project: "When the temps weren't needed anymore, it was common for groups of them to be rounded up, summarily let go without notice, and told that a call would be forthcoming if their services were needed again."[42]

Other aspects are typical of the sector as a whole, such as the phenomenon of "crunch time." QA testers' very low salaries constitute the bottom rung of Smalline's salary structure, but these workers share many problems with other teams working in the same offices. The gaming sector is famous for the number of overtime hours its employees are expected to perform. A paradigmatic case is "crunch time"—the days and hours before a game, a new level, or an important update is released. One of Smalline's workers describes it as follows:

> When crunch time is called, people bring their sleeping bags and stay for five days to finish the release. You cannot leave the office; you cannot go home. We order pizza and work all the time. If there are people who want to go home to look after their kids, they face negative sentiments.[43]

This scenario hints at one reason why the gaming workforce has remained predominantly male. Caring for children, still predominantly the burden of women, does not typically allow for long and hyperflexible working hours. The Smalline works council cites not only working hours but also a widespread sexism in the gaming sector—exemplified, for example, by #gamergate, a high-profile scandal around sexism in gaming—as a reason for the low proportion of female workers. In fact, these interrelated issues, labor intensification especially through crunch time, and the struggles around the rampant sexism in gaming culture prove to be two of the most important points of conflict and contestation in the global gaming industry.[44] Crunch time is an old and well-known fact in gaming. It seems that the neoliberal ethos of the "New Economy" combined with industry professionalization fosters harsh working conditions that tend to combine the downsides of both work cultures—professional and creative—for employees. The start-up culture of the New Economy has always combined a "creative" and "free" working culture with extensive and often uncompensated overtime. Particularly in the big studios, however, the flat hierarchies and free spirit have vanished, while the overtime has remained.

THE CASE OF ELECTRONIC ARTS

Today, the global gaming industry generates yearly revenues well over $150 billion. The record-breaking game Grand Theft Auto V, released in 2013, for example, has passed $6 billion in sales (more than double the highest-grossing movie of all time as of this writing, *Avengers: Endgame*). The industry centers on big studios such as Rockstar, the publisher of Grand Theft Auto; Tencent; Ubisoft; Sony; and Electronic Arts (EA). EA employs around 9,000 workers, 1,300 of whom work in its largest facility in Vancouver. EA fashions this digital factory as a campus, replete with gyms, gourmet cuisine, cultural programming, and so on—quite similar to the Googleplex in California. Across all digital factories, amenities such as gyms and basketball courts are key to fashioning a creative and free work culture. However, these facilities tend to be suspiciously empty whenever visits are permitted. A Smalline employee answered my question regarding their vacant table tennis table by explaining that "sometimes we use it for a two-person meeting. You can play and talk about the project. That's good for your back."[45]

The gaming industry still lives off the myth of work as play, stemming from the early days of video games. Many of the first computer games were designed out of boredom and not taken seriously as commercial or cultural objects. Mostly produced in academic computing departments where universities cooperated with the military in developing weapons for the Cold War, these games were often spinoffs of simulation technologies and largely distractions for the scientists themselves. As with the computer and even more so the internet, the nexus between the US military and universities established during the Cold War was a central driving force behind the development of digital games.[46] For decades, games were mostly by-products of and distractions for the scientists and engineers employed in these labs, as the famous game Spacewar developed at Massachusetts Institute of Technology in 1962. Their Russian opponents, though less productive in creating such distractions, can at least claim the invention of the still hugely popular game Tetris by a computer scientist at Dorodnitsyn Computing Centre of the Academy of Science of the USSR in Moscow.

North American students from some of those labs began to circulate these games. Spacewar was circulated by computer scientists via the internet's precursor, the military's ARPANET.[47] In the context of the Vietnam War and the student protest movement, some of the younger scientists grew increasingly critical of the state and military sponsoring their projects. In this sense, the development and playing of games during working hours can also be understood as a form of low-level protest; in those years, "virtual games were a refusal of work: they signified leisure, hedonism, and irresponsibility against the clock punching, discipline, and productivity," as Greig De Peuter and Nick Dyer-Witheford note in their seminal book *Games of Empire*.[48] They cite especially Atari, an early gaming company that embodied a free-spirited "antiwork" culture in its early days in the 1970s. With regard to today's bigger gaming corporations, they argue, however, that "this anarchic self-image, a hangover from Atari days, though perhaps still somewhat true of small game companies, hardly stands up to an encounter with a behemoth like EA—yet it remains a mythic element in the allure of game work."[49]

In 2004, EA found itself at the center of a controversy over working conditions in the gaming industry after a public letter written by the partner of an EA employee provoked a scandal. The blog post signed "EA spouse" complained about her partner's exploitation in bitter tones, finally asking EA's then-CEO, Larry Probst, "You do realize what you're doing to your people, right?"[50] The blog post triggered an ongoing debate in the gaming industry concerning the conditions of employment. A crucial point of the 2004 open letter to EA was crunch time, described not as an exception but as normal in many production circuits: "Every step of the way, the project remained on schedule. Crunching neither accelerated this nor slowed it down; its effect on the actual product was not measurable. The extended hours were deliberate and planned; the management knew what it was doing as it did it."[51]

EA is an example of how the gaming industry has professionalized while continuing to use the sector's anarchic image to raise working hours and pressure on workers. Concerning the 2004 conflict, the *Wall Street Journal* reported that "far from the hip, creative image Electronic Arts conveys, work inside the company more resembles a fast-moving, round-the-clock auto assembly line."[52] In 2005 and 2006, EA settled class-action lawsuits over unpaid overtime with software engineers for $14.9 million and graphic artists for $15.6 million. While EA workers have since won significant improvements, the overall picture remains unchanged in many aspects. Most workers feel that crunch is still expected from them. This suggests that crunch time is a regular strategy deployed to exploit digital labor and avoid labor regulations. In many instances, crunch time appears to be a normal strategy to appropriate more hours from the workforce. The "free spirit" and "passion" of creative labor are crucial cultural factors that enable these strategies.

Even if crunch time remains an important component of the organization of labor and increasing surplus value, the industry has come a long way from its origins in engineers' garages and basements. Smalline's Berlin offices possess all the artifacts one would expect for a gaming company; nevertheless, at many points it is difficult to differentiate them from other IT offices in the vicinity. The gaming sector has witnessed explosive growth alongside severe crises in recent years, giving many workers the impression that the sector "has grown up."

Game designers and programmers form the core team among employees, and they have better conditions and higher salaries than the testers. However, the hacker image upheld in some sections of the software and gaming industry tends to obscure forms of standardization and routinization that often characterize programmers' work as well. Code "is labor crystallized in software form," and although the terms under which it is produced vary widely, the labor of coding tends to be quite repetitive and boring.[53] Media theorist Jussi Parikka has pointed to a long history of the "dull" and laborious site of programming. From Alan Turing's remark that computing's future was in the "office work of programming," to the division of coding labor into creative "metaprogramming" and its technical execution at PARC Xerox in the 1970s up to the present, Parikka sketches a cultural history of "software work as factory work."[54] The gaming sector is by no means immune to these tendencies, even in the more prestigious and creative areas of coding, which are increasingly subject to standardization, decomposition, and quotas.

A study on workers in the Australian gaming industry identifies similar developments: namely, a trend toward professionalization, at least among larger studios, which entails more permanent contracts and an increase of structure, standards, and hierarchies in the production process.[55] This tendency also leads to decomposition and a more sophisticated division of labor, fostering specialization and more repetitive tasks. One worker is quoted as stating: "I opened the same door like five million times and I heard the same sound effects five thousand times. Play ViroShot every single day for the rest of this year and I bet on day 365 it won't be the most brilliant game you've ever played." Another respondent adds: "You've got a guy who creates heads and you've got a guy who creates bodies and you got a guy who makes environments and you've got a guy who animates them."[56] Outsourcing is a big part of this process, as in Berlin. "A lot of the graphic work is outsourced to big graphic sweatshops in South Korea, which are specialized to manufacture three hundred pieces of digital armor a day, with various textures, automatic exposure, gamma correction, and 3D models," explains a Smalline employee.[57] With this, this labor moves precisely to the region where the business of professional gold farming took its first steps.

CONFLICT, PLEASURE, MATERIALITY

The conflicts of the past few decades have contributed to a professionalization of the industry, the benefits of which are unevenly distributed. Gaming labor is stratified: some workers enjoy relative security and higher pay, while many of the risks are placed on more flexible and precarious workers. The examples of this chapter show how especially the frictions between these labor regimes can be the source of discontent and labor struggles. Recent years have seen different sites and issues of conflict in the international video games industry. Newer initiatives such as Game Workers Unite and Le Syndicat des Travailleurs et Travailleuses du Jeu Vidéo show how the desire to organize and find means of collective expression continues to spread among video game workers.[58]

Clearly, no linear and frictionless process of the Taylorization of creative work can be found in the gaming sector. Instead, this sector shows a complex recombination of different labor regimes that both have a history in the production of games. The pleasures of play as work are sometimes real, and even the digital workers in Chinese gold farms sometimes find enjoyment in their work. Many workers in gaming are also video game fans themselves, and their pleasure at participating in the industry is as real as the pain caused by long hours and low pay. To dismiss the allure and culture of creativity as a simple ideology designed to better exploit labor falls short of the mark. Rather, it is necessary to understand the relation and development of this labor regime in the context of rationalization and a deepening and more complex division of labor.

Even if the work processes of Berlin-based testers and Chinese farmers are oftentimes somewhat similar, their general conditions vary tremendously. In general, the gaming industry involves manifold and varied circuits of production and circulation, from coltan mining in the Democratic Republic of Congo, computer assembly in Mexico, coding labor in California, narrative development in Berlin, graphical development in the US, and the digital laborers producing millions of pixelated objects in South Korea, customer service hotlines in India or quality assurance in Berlin, to CD and game packaging production in China—even the production of a single game connects a wide variety of economic circuits and layers. The digital workforce is distributed all over the globe, segmented, and hierarchized in various ways.

This list illustrates, finally, the very materiality of the labor, infrastructure, and products of gaming. Online multiplayer games produce a massive quantity of data. Both the huge amount of data and the advantages both leisure and professional players can access through faster internet connections demonstrate the importance of personal computers, routers, network connections and data centers, fiber-optic cables, and the rest of digital connectivity's very material infrastructure. These infrastructures and their need for electricity express another side of gaming's materiality: in 2006, Nicholas Carr estimated that an avatar in the game *Second Life* consumes as much electricity as the average citizen of Brazil.[59]

These computer and internet infrastructures interact with national borders, the specific local cost of the reproduction of labor, language skills, payment infrastructures, and many more factors to produce a complex economic cartography. To conceptualize this cartography, both the conceptual apparatus and interaction of national economies and an understanding of the internet as a flat and borderless plane fall short of the mark. Rather, it is necessary to understand that these global infrastructural connections and disconnections imply a multiplication and fragmentation of borders and, with this, a multiplication and fragmentation of figures of labor and migration.

4

THE DISTRIBUTED FACTORY

Crowdwork

In 2013, artist and architect Nick Masterton produced the short video clip *Outsourcing Offshore*.[1] In this case, however, *produced* meant that, instead of generating the video's content himself, he outsourced all photography and narration to a pool of distributed online workers on so-called crowdwork platforms such as Amazon Mechanical Turk, Task Rabbit, and Fiverr. On these platforms, employers post small tasks that workers can perform on their home computers for a small fee. Masterton created tasks asking workers to photograph their workplaces or lunches or record short audio files about their work lives. The workers were asked about their mode of transportation, to sing their favorite songs, or talk about their hopes and concerns for the future. The artist then edited the results of the tasks together into a short video clip replete with audiovisual impressions of workers' lives across the globe, as they sit in front of their personal computers and solve a wide array of tasks brought to them by crowdwork platforms.

The intimacy facilitated by the workers' intermingling voices and stories is complemented by the photos of the workplaces, which are mostly personal computers in the worker's homes, in bedrooms, at dinner tables, or terraces. Shown in a montage, these pictures reveal a scattered factory in which thousands of workers in very different situations labor in isolation from one another. The platforms' algorithmic architectures articulate these workers, organize their invisible cooperation, and set them into global competition. The abstraction and intimacy of *Outsourcing Offshore* aptly capture the digital factory of the platform—a distributed bedroom factory articulating a global, heterogeneous workforce in a seemingly weightless way.

Today, platforms such as those used by Masterton for his video project employ millions of digital workers from all over the globe. Working from their personal computers, they constitute a hyperflexible, on-demand workforce that can be accessed and let go in seconds. Most of them sweat over minor tasks that are not (yet) computable by machines but can easily be solved by a distributed mass of human cognition organized by algorithmic infrastructures. Online labor platforms enact new forms of control and flexibility and serve as decentralized sites of digital production that are crucial to many nodes of the global economy, most notably the production and training of artificial intelligence (AI). Today, millions of workers are logged into digital labor platforms to categorize pictures, test software, transcribe audio recordings, or optimize search engine results. Often hidden from view and dispersed around the globe, these workers nonetheless form a growing component of the digital working class as well as the political economy of the internet more generally.

In the context of this book, the digital piecework organized through crowdwork platforms represents both a paradigmatic and specific instantiation of the digital factory. In crowdwork, digital technology allows for new modes of standardization, decomposition, quantification, and surveillance of labor—often through forms of (semi-)automated management, cooperation, and control. Crowdwork platforms (similar to many other gig economy platforms) are characterized by algorithmic management and automated surveillance. The workers of these platforms are independent contractors whose contractual relationships with the platform last only as long as they take for the completion of these small tasks, often a matter of seconds. This form of on-demand labor is therefore characterized by radical flexibility and contingency. In this respect, digital platforms stand in a centuries-old tradition of contingent work, including things such as migrant day labor and home-based piecework.

New forms of digital control and old forms of contractual flexibility, in combination with the platform's ability to potentially reach all people connected to the internet, allow for the inclusion of the deeply heterogeneous and globally dispersed workforce profiled in Masterton's *Outsourcing Online*. Unlike a traditional factory, the platform needs a low degree of homogenization in terms of space, time, or lifestyle. With this, an important distinction vis-à-vis classical Taylorism becomes evident: namely, digital Taylorism does not produce a digital "mass worker" in the Fordist sense. Among other things, this chapter analyses the multiplication of labor by way of describing how crowdwork allows the generation and exploitation of a new pool of workers, disproportionately comprised of women and people with care responsibilities, and new digital workers in the (rural) Global South. Digital platforms allow companies to access a very diverse global pool of workers who are available on demand and can be let go in seconds. A forerunner of this form of labor was—once more—Amazon.

"PEOPLE-AS-A-SERVICE"

In a speech held at the Massachusetts Institute of Technology (MIT) in 2006, Amazon founder and CEO Jeff Bezos promised insights into the "hidden Amazon"—the parts of the company lesser known than the world-famous online retail platform.[2] He presented the services operating under the name Amazon Web Services (AWS) as several online offerings, particularly cloud computing infrastructure. Although AWS attracts little public interest, it has made Amazon the most important provider of cloud computing services on a global scale, with customers ranging from the streaming platform Netflix to the Central Intelligence Agency. Bezos, however, opened his 2006 presentation at MIT not with the server farms and data cables but, rather, another branch of AWS: its crowdwork platform, Amazon Mechanical Turk. The principle of this platform providing living labor as a service follows the logic of cloud computing: outsourced, flexible, scalable, available on demand.

Like other AWS branches, the Mechanical Turk was first developed as a solution to a problem Amazon encountered in its business operations. In constructing its online marketplace, Amazon had attempted to develop software capable of reliably recognizing all duplicate and inappropriate products on the site. However, the task proved to be noncomputerizable. Rather than hire additional workers to assist the software, Amazon developed the Mechanical Turk platform to outsource the work as human intelligence tasks (HITs) to the crowd of internet users. The software generated a preselected list of products and uploaded the presumed duplicates to the Mechanical Turk platform. Workers registered there then determined whether the items truly were duplicates and were paid two cents in return. This principle of platform-based outsourcing of noncomputerizable tasks to the flexible and scalable workforce of the crowd proved to be a winning model, and Amazon soon opened its platform to other companies to outsource their work online for a fee. Soon other corporations began to outsource small tasks to the crowd using Mechanical Turk, including categorization of pictures, correction of spelling errors in texts, product descriptions, searching for email addresses, participation in different surveys, digitalization, and categorization of all kinds of data.

Hiring workers for data-processing tasks is nothing new as such. Clerks, secretaries, and telegraph messengers have been around for centuries. More recent examples include call centers, which often hire homeworkers for various jobs and people who perform translations and other jobs from home. The early internet economy itself was fueled by paid and unpaid labor performed by people working from home outside their regular jobs from the outset: chatroom moderators, software testers, hobby developers and game modders, mailing list participants, and many more.[3] The speed, size, and algorithmic organization enabled by crowdwork platforms, however, makes for a new, unique quality of crowd labor: "Instead of hiring hundreds of homeworkers for a few weeks, a single person can hire sixty thousand workers for two days. This shift in speed and scale produces a qualitative change in which human workers come to be understood as computation," explains Lilly Irani, who has done groundbreaking research on human-computer interaction and Amazon Mechanical Turk.[4]

Other platforms soon started to copy the business models of Amazon's forerunner platform, and today Mechanical Turk is only one of the thousands of crowdwork platforms operating according to the principle Bezos formulated aptly at MIT in 2006: "You've heard of software-as-a-service. Well, this is basically people-as-a-service."[5]

A GLOBAL ECOLOGY OF ON-DEMAND LABOR

Labor organized via digital platforms is a growing global phenomenon. The logic by which crowdwork platforms organize and control labor is characteristic of the gig economy, which continues to broaden its scope into ever new sectors. Labor on platforms such as Uber, Helpling, or Deliveroo is executed "offline" and hence bound to specific localities; apart from this, however, these platforms express a similar logic of on-demand labor, flexible contracts, and automated management. More generally, the labor relations that underlie crowdwork platforms for digital labor are to be understood within the rise of forms of algorithmic management and surveillance and the ongoing flexibilization of labor markets throughout different sectors. These developments are broad and multifaceted, reaching far beyond the confines of the gig economy and digital platforms. Many of the logics described here are in that sense not limited to digital labor platforms but must be analyzed as an important tendency of the transformation of labor in digital capitalism. Hence, while crowdwork platforms are clearly a specific and, in some aspects, extreme example of algorithmically controlled on-demand labor, they are also an expression of broader tendencies within the world of work and worthy of careful analysis as laboratories for the future of work.

The term *crowdwork* broadly defines labor that is outsourced via an online platform to a large group of people working remotely via their digital devices. Accordingly, crowdwork platforms are mediating *digital labor*—here pragmatically defined as labor that centrally involves the manipulation of data by means of a digital device such as a laptop or smartphone. This is the central difference to location-based gig economy platforms like Uber or Deliveroo: here, workers use cars, scooters, and bikes to transport passengers and food through cities. Just like these workers, however, crowdworkers are not employees of the platforms but are typically described as "independent contractors"—freelance workers without regular contracts. Most crowdwork platforms function as intermediaries, allowing other corporations to outsource a great variety of tasks to a global pool of on-demand workers in exchange for a fee. Platforms typically not only organize the mediation of these tasks but also control the labor process in minute detail, handle payment and ratings, and so on; that said, they tend to strategically position themselves simply as online labor markets or tech companies in order to avoid responsibilities to their workers.

Measuring the size of the crowdworking class is a difficult and complex task due to its irregular, often informal, and very global nature. Since the early days of Mechanical Turk, the business has exploded in size and diversified considerably, and today, Amazon's forerunner platform is outsized by far by many of its competitors. Internationally, the largest platforms include Freelancer.com, a platform that offers a wide range of digital work and counted over twenty-seven million registered freelancers in 2019, and Upwork.com, with over twelve million workers. Traditional labor market statistics often fail to account for these forms of labor, and attempts at quantifying the overall size of this labor market are to be understood as rough estimations. In 2015, the World Bank estimated the number of registered workers on these platforms at around forty-eight million; another estimate from 2017 that takes Chinese platforms into account puts the number at seventy million registered workers.[6] Of course, not all workers are active all the time, and a portion of these accounts may be dormant, so the number of active crowdworkers will be considerably lower, though still a double-digit million amount. Taken together, these studies indicate that platform labor is starting to become an important factor in labor markets, in both the Global South and the Global North.

Today, crowdwork platforms for online labor exist in very different forms and purposes. Some are platforms for more complex tasks (e.g., for coders, designers, translators). On these platforms, freelance workers compete globally for work, sometimes in the form of tenders and competitions. In the world of these platforms, rankings, qualifications, and experience are important metrics, and the profiles of workers are crucial for them to secure further jobs. Platforms such as these platforms, which often organize highly qualified and complex work, mark one end of the spectrum of crowdwork platforms, often referred to as "macrowork." Platforms such as Amazon Mechanical Turk, in contrast, provide mostly simple tasks and mark the other of the spectrum of crowdwork. The majority of these tasks are repetitive and take little time to complete. The investigation of this chapter focuses on this end of the spectrum, described as "microwork" by Jeff Bezos in his 2006 speech at MIT: "Think of it as microwork, so for a penny, you might pay someone to tell you if there is a human in a photo."[7]

"Access a global, on-demand, 24×7 workforce," advertises Amazon Mechanical Turk today. "MTurk is well-suited to take on simple and repetitive tasks in your workflows which need to be handled manually," the website claims.[8] For a worker logged in, this translates most of the time into simple tasks that can be solved in minutes or even seconds. A random look at Mechanical Turk reveals the following HITs: "Looking at a receipt image, identify the business of the receipt—reward $0.02"; "Type the text from the images, carefully—reward $0.01"; or "Provide as many tags as possible for an image in the photo—reward $0.02."[9] Several hundred thousand such tasks are available on the microtask platform.

Typically, these small, simple tasks are paid by the piece and require no formal qualification. They include things such as categorization of pictures, transcription of speech, product descriptions, recording of pictures or small videos, participation in surveys, digitalization, and categorization of various data. While this chapter concentrates on this end of the spectrum of crowdwork, a clear distinction between micro- and macrowork is hard to draw, as the nature of tasks varies greatly even on one given platform. In any case, it is important to bear in mind that today crowdwork is not limited to menial data processing but rather continues to eat its way into other sectors of digital labor.

THE LABOR BEHIND AI

In recent years, the training and optimization of AI have become the main factor pushing the dynamics of crowdwork. Among other things, the development of AI is based on huge categorized training data sets, the production of which necessitates large amounts of human labor. Today, crowdwork platforms are providing millions of hours of hidden labor that is necessary for algorithms that power self-driving cars or allow devices to understand human language. Clickworker, a German platform with over 1.8 million registered workers located in over 130 countries, advertises its services specifically to developers of machine-learning software: "Improve your AI systems and algorithms with training data that is optimized by humans for machine learning. Our Clickworkers handle projects of all sizes, helping you train your AI systems, improve search relevancy, and increase the overall efficiency of your core services."[10]

While Clickworker also offers a range of other services and tasks from search engine optimization to content creation, other platforms have focused their business solely on the booming sector of training data for AI applications. CrowdFlower, a very early crowdwork company, rebranded itself as Figure 8 in 2018 and started to focus exclusively on training data sets for machine-learning applications. This platform, in turn, was acquired by Appen in 2019, another crowdwork platform focusing on AI. "Training data isn't labeled or collected on its own. Human intelligence is required to create and annotate reliable training data," advertises Appen on its website. "Our platform collects and labels images, text, speech, audio, video, and sensor data to help you build, train, and continuously improve the most innovative artificial intelligence systems."[11] Like the over one million workers of this platform, an increasing number of digital workers across many different platforms are involved in the development, training, and support of AI training speech recognition software for smart home applications, recording videos of hand gestures to train digital assistants, or mark pedestrians and traffic lights in photos to train algorithms for autonomous driving.

While all these examples and many more are crucial fields in which massive amounts of money are invested into AI systems, the data-hungry field of autonomous driving is especially important to crowdwork. Recent years have been characterized by enormous investments into the ambitious development of fully autonomous vehicles (and more modest assistance systems), both by traditional automotive producers like Ford, Volkswagen, and General Motors and lateral entrants like Uber, Apple, and Google, whose subsidiary Waymo is by many accounts the current leader in development. The race to the market for autonomous driving is characterized by massive capital investments, broad public debate, and harsh competition between companies and conglomerates, including standoffs such as a major lawsuit over the stealing of trade secrets between Uber and Google. A crucial component in this competition for the development of autonomous cars are annotated data sets with pictures of all kinds of things a car encounters in traffic.

As security is crucial, driverless cars must be able to recognize everything they come across in all possible situations, including other cars, pedestrians, cyclists, traffic lights, police traffic checks, animals, construction sites, and potholes. To train algorithms, developers need a massive amount of annotated photo and video material. This is why workers all around the globe logged into crowdwork platforms spend the better part of their days marking and labeling objects in video and photo material. The material is mostly taken from videos shot in traffic, and the tasks involve various forms of labeling the different objects in the shots. Other workers double-check the labeling of their colleagues or the decisions various algorithms have taken in simulations. The amount of training data needed as well as new requirements concerning precision have not only caused many existing platforms to concentrate on data annotation for machine learning but helped create new platforms specifically catering to the needs of developers of software for autonomous cars (e.g., Scale, Mighty AI).[12] The platform Mighty AI (whose worker-facing interface is called Spare 5) was acquired by Uber in 2019 and closed its business for other customers to exclusively serve Uber's development of self-driving cars.

Just as in many other sectors where labor outsourced through crowdwork platforms plays an important role, the hidden contribution of crowdworkers to autonomous driving systems exemplifies how high-profile processes of automation are massively fueled by human labor. The "ghost work" of hundreds of thousands of platform workers annotating data is crucial for the development of self-driving cars.[13] As in many cases of automation, the discourse around autonomous driving is focused on technology and tends to overstate the ability of algorithms while obscuring the human labor necessary.

The importance of human labor power for supposedly automated systems is reflected well by Amazon's forerunner platform name: "Mechanical Turk" refers to a chess computer that attracted considerable attention in the eighteenth century. The apparatus consisted of a "Turkish" puppet and an apparently sophisticated apparatus. The chess computer was surprisingly successful and even won against Napoleon Bonaparte, among many others, according to legend. However, its secret was simple: the Turk was in fact a mechanical illusion that allowed a very skilled human hiding inside to operate it.[14] Referenced both by Alan Turing, ironically as an early example for research into AI, as well as Walter Benjamin, as an allegory for historical materialism and its relation to theology, the Mechanical Turk has become a symbol for the borderlands of human and machine, science and magic. Amazon's decision to name its platform after this "first computer" is telling, as is the slogan the company used to advertise for the platform in its early years: "Artificial Artificial Intelligence."

The development of AI and its employment in complex settings has been a very dynamic process, but it has also been characterized by major setbacks and the continued importance of human labor. In fact, after a period of enthusiasm, the development of self-driving cars has slowed in recent years. The shortcomings of autonomous cars in dealing with the endless multitude of complex situations that characterize traffic outside controlled and limited test zones have boosted skeptical voices claiming that fully autonomous cars will not be a reality in the foreseeable future. Some companies have suggested that a possible way to help autonomous vehicles stuck in complex situations could be workers on call in remote locations. These so-called teleoperators working in call centers or on-demand via platforms could be called on by the car's software to its cameras and help navigate the situation. This scenario shows a possible future occupation for crowdworkers: navigating stuck autonomous cars from their bedrooms and kitchens.

In any case, the present situation of global labor on digital platforms shows that automation cannot currently be understood as the simple elimination of jobs by increasingly smart software and sophisticated robots. Rather, crowdwork is a major example of how the development of new technology shakes the world of labor, destroys some jobs but also creates new ones, and changes value chains in the process. Jobs that disappear at one point often reappear in changed forms at another point. In the case of crowdwork, labor that is supposedly done by algorithms is often actually done, or at least supported by, a multitude of on-demand workers hidden in private homes in Germany, internet cafés in Venezuela, or the streets of Kenya. These workers have a complex relationship with the progress of machine learning; they are important drivers of its development, but they are constantly in danger of being replaced by these products. As mentioned previously, this development does, however, create ever new demands for human labor. With a view to these workers, it is less important to speculate about a future without work than to analyze a present characterized by the emergence of a truly global market for digital labor, new forms of organizing and controlling remote workers, as well as new forms of resistance.

"I NEED TO MAKE €100 TO MAKE ENDS MEET"

Daniel is a crowdworker. The twenty-seven-year-old read about the industry in a newspaper and decided to try out various platforms. Today, he works mainly for CrowdGuru, a German platform with a comparatively small crowd of around fifty thousand. His desk in a student flat in Berlin's Wedding district is dominated by two screens and an hourglass. He needs this job. His parents support his studies at the Technical University of Berlin, where he also works as a student assistant, but the money is not enough. "I need to earn €100 per month to make ends meet," he explains, "and most of the time it works, in my best month I even made more than €400."[15] Rents are rising rapidly, even in the run-down old working-class neighborhood of Wedding, and there are few options to save money. Crowdworking also allows him small luxuries like buying books.

Daniel has specialized in small texting jobs. Besides classical microtasks, the platform CrowdGuru also offers a few texting jobs. These jobs are mostly product descriptions and small advertising texts for online shops and other companies. Original texts featuring the proper keywords are very important for websites to appear at the top of search engine results; this has resulted in a dire need for texts written by humans, which is met by crowdworkers like Daniel. Most texts are rather short; a typical task is a product description, perhaps for the webshop of a hardware store, some two hundred words long, which will pay one or two euro. Daniel has become a specialist for these jobs. Most of his texts are product descriptions for online shops, especially hardware stores and furniture shops: "I've found a thousand ways to describe curtains," he laughs.[16]

To Daniel, crowdworking's biggest advantage is the flexible working hours. Given his studies and second job, it would be nearly impossible to find another job with fixed hours. Crowdworking can be done in between other tasks, whenever he wants: "Food in the oven—half an hour of working; if there is a break between two lectures, I'll quickly write another text on curtains on my laptop."[17] Next to the computer screen on his desk sits a large hourglass. "I use it to measure whether I can write a text in a time that is worthwhile," says Daniel. For him, *worthwhile* means an hourly wage over five or six euros (a number well below official minimum wages in Germany). When there are enough texting jobs, he is able to reach his goal. However, his biggest problem is the fluctuating number of jobs available on the platform. Sometimes, no lucrative jobs can be found, which means financial problems for Daniel who has no savings. This situation leads to a very common pattern among crowdworkers: the platform is always kept open in a browser tab and regularly checked for jobs. Competition is tight, and well-paying jobs can disappear within minutes. If no texting jobs are available, Daniel has to turn to more classical microjobs like data processing or photo tagging, even if he dislikes this form of crowdwork: "This digital assembly-line stuff doesn't make sense for me. You need to be extremely fast to make more than three euro an hour."

THE DIGITAL ASSEMBLY LINE

His notion of the "digital assembly line" is a very fitting description for a majority of microtasks. Labor on crowdwork platforms organizing such jobs is most of the time radically decomposed and standardized. A great part of the work consists of huge data sets that are decomposed into microtasks, very small jobs that can be solved in minutes or even seconds. For this to work and to be profitable, the organization of labor and the cooperation between a huge number of crowdworkers must be organized mostly automatically. This division of labor and form of cooperation works out of sight of the workers and is orchestrated automatically by the platform, a form of algorithmically organized cooperation. Daniel is cooperating with a great number of workers from all over the globe without knowing it. In this sense even if working on a platform can make him feel like the only worker in a deserted factory, the nature of crowdwork is often highly cooperative.

As a result, for a crowdworker like Daniel, it is often hard to know the meaning of one's labor while performing these tasks. For example, one can only guess at the logic behind checking Ryanair's website for the price of a specific flight countless times for five cents each. Often, crowdworkers have no clue of the exact purpose of the task at hand. For example, Project Maven, the controversial Pentagon-sponsored AI warfare project using machine learning to distinguish people and objects in drone videos, involved not only engineers from Google, many of whom ended up revolting against the project, but also digital workers from a crowdwork platform. The only difference is that these workers could not protest against the project because they did not know they were involved in it. Crowdworkers registered on the platform CrowdFlower/Figure 8 were used to label objects in satellite images to train the software without knowing that they worked for Google or the Pentagon.[18]

Microwork is typically highly standardized; various technologies of algorithmic tracking, tracing, and rating of the labor of the individual crowdworkers are in place. Some platforms allow clients to control workers via random screenshots or keystroke counters. Often the platform will decide automatically or let the customers decide if a task is done successfully and rate workers accordingly. Amazon Mechanical Turk exhibits the most radical regulations in this respect: here, purchasers can judge whether a task is solved convincingly and thus if they are required to pay. The rights to the product remain with the purchaser, whether remunerated or not. This system leads to situations in which workers feel their work is rejected unfairly. These workers depend on positive ratings from the task requesters to receive more work in the future. Accordingly, for them, a rejected task means not only potential loss of payment, but potentially restricted access to further work.

Complaining about nonpayment or other issues is often quite

difficult. The technical design of many platforms does not allow for direct communication between requesters and workers, and some even ban their workers from trying to contact requesters outright. This is one of the biggest sources of discontent among crowdworkers, as they are unable to discuss issues such as arbitrary nonpayment or logical errors in the tasks with the requesters who issue them. Protesting against these conditions, workers have used the slogan "we are no robots," a slogan notably also used by workers in Amazon distribution centers.

While some platforms spend time and effort to communicate with their workers and develop a community of workers with deeper ties to a specific platform, in microtasking, human labor is typically managed algorithmically. This is part of their logic to integrate human labor power into computing infrastructure as seamlessly as possible. Importantly, it is also a question of costs. As one large-scale requester explained to researcher Lilly Irani: "You cannot spend time exchanging e-mail [with the workers]. The time you spent looking at the e-mail costs more than what you paid them. This has to function on autopilot as an algorithmic system."[19]

A central function of crowdwork is the automated insertion of human labor into complex algorithmic architectures. Many platforms are designed to allow automated access to a distributed pool of human workers. Through an application programming interface (API), it is possible for the software to automatically create a task on a platform should it require the aid of human cognition. One such example might be an algorithm searching social media for offensive content to be deleted. Such an algorithm might automatically sort through pictures uploaded to a social media platform to delete, for example, pictures containing nudity. If the software is unsure, it can be programmed to automatically upload the picture onto a crowdwork platform, pay two cents to a crowdworker like Daniel who decides whether the picture is offensive or not, automatically compute the answer, and move on (and potentially even learn from it).

This process hints again at the importance of the labor of crowdworkers for AI, both in the development of such algorithms and in assisting them. Countless instances similar to the preceding example occur in which software can perform very complex tasks but is hampered by problems that humans can easily solve. These roadblocks are often cultural, contextual, or graphical problems that are difficult to compute by software. Although software develops rapidly, the number of sites across the political economy of the internet and beyond where human labor is necessary because software lacks the cultural or contextual knowledge or graphic and audio skills remains vast. Crowdwork platforms address these gaps in computing and provide flexible and scalable labor power to fill them. This labor, however, is often masked as technology and remains hidden behind the screen.

ON-DEMAND LABOR

From capital's perspective, crowdwork allows the creation of a hyperflexible, scalable, on-demand workforce that can be accessed and let go instantly, leaving employers with little responsibility. For IT companies outsourcing menial work through these platforms, crowdworking allows for experimentation with different forms of human labor, while also allowing said companies to portray themselves as technology companies rather than labor companies—a strategic move often key to attracting venture capital.[20]

Crowdworking platforms often claim to be mere intermediaries between employers and employees, somewhat resembling online labor markets. Upon closer examination of the platforms and how they structure the labor process, however, it becomes evident they are anything but a neutral intermediary between labor and capital. As infrastructures for digital labor, they assume many of the socio-spatial functions of the traditional factory. Just as in Amazon's distribution centers, labor on crowdwork platforms is characterized by a high degree of control, standardization, and decomposition. Most platforms have precise technologies for the measurement and surveillance of labor. The majority of the processes work automatically, which increases the information asymmetries and forms of unidirectional command between platforms and workers. It is precisely the standardization of tasks, the means of algorithmic management, and surveillance to organize the labor process, as well as the automated measuring of results and feedback that are key characteristics of what is described herein as digital Taylorism.

While the possibilities of digital technology organizing distributed workers by means of decomposition, standardization, and surveillance represent one crucial aspect of crowdwork, radical flexibility enabled through specific contractual and wage forms represent another. In crowdwork, we find yet again the specific combination of new forms of algorithmic management and digital control with flexible and contingent labor arrangements. It is the very possibility of digital technology allowing the precise organization, measurement, and control of living labor that in turn enables the flexibilization and multiplication of labor. In the case of crowdwork, it allows including a very heterogeneous and distributed group of workers into a precisely organized digital factory.

Hereby, it becomes evident that crowdwork cannot be understood as a simple rebirth of Taylorism, because the way labor is organized in spatial and contractual terms in fact tends toward the exact opposite of the Taylorist factory. Besides the fact that Daniel and his colleagues are dispersed around the globe, the legal form of their employment is another crucial difference from most traditional factories. The flexible forms of incorporating workers on demand into this factory are a crucial characteristic of crowdwork and an important difference from Taylorist factories. The contractual modes of regulating workers are designed to foster maximum flexibility and free the platforms of any obligations vis-à-vis their workers. Tendencies such as the legal construct of the independent contractor and the return of piece wages as discussed in previous chapters find what is probably their most radical expression on crowdwork platforms. These elements prove crucial to digital Taylorism—together with the digital modes of organizing the labor process, they not only help create a scalable and hyperflexible workforce but are also crucial to securing the subsumption of labor under capital outside the disciplinary space of the factory.

Most platforms consider their workers "independent contractors" who are—especially in microwork—paid by the task. Logged onto a platform, the legal relationship of the crowdworkers to the purchaser of their labor power only lasts for the length of the task. On microwork platforms, this often only lasts a few seconds or minutes. Logged onto such a platform, a worker might find several photo tagging tasks. Upon accepting a task, they might be asked if there is a human in the photo. The question is answered, and the platform transfers a few cents on the account of the worker, who is already onto the next task.

The definition of the workers as independent contractors is central to the labor relations of most gig economy platforms, shifting the relationship outside the realm of many labor laws and regulations designed for standard employment. As specified in the terms and conditions of the platforms, independent contractors are not entitled to the benefits many regular employees enjoy, such as vacation pay, sick leave, paternity leave, insurance programs, or unemployment benefits. Furthermore, with these tasks paid by the piece we see the emergence of a seemingly outdated wage relation: the piece wage.

Karl Marx once described piece wages as "the form of wage most appropriate to the capitalist mode of production."[21] Although more common at the time of *Capital*'s writing (especially in the cottage industry organized on the putting-out model) and subsequently marginalized over the history of capitalism, Marx's characterization of the piece wage still proves surprisingly helpful in understanding its (contractual) function in the world of online labor platforms and beyond. The piece wage system, as well as the digital possibilities of measuring and controlling the product of work, removes the need to control and supervise the worker: "Since the quality and intensity of the work are here controlled by the very form of the wage, superintendence of labour becomes to a great extent superfluous," as Marx puts it.[22] Particularly in microtasking, the speed and amount of time in which work is performed are determined by the worker. However, the duration and intensity of labor are directly reflected in the amount of income a crowdworker can generate. In this sense, the conflict between the factory owner and supervisors on one side and workers on the other over the intensity and duration of labor is displaced to the worker. Marx shows how hourly wages and piece wages have historically coexisted, sometimes even in one factory. The particularities of the piece wage as a form of labor intensification as well as a form of displacement of control of the work process are crucial components of crowdwork.

The legal construct of the independent contractor is closely bound with the return of the piece wage in digital capitalism, as can be seen across different sectors, from harbor truckers and bicycle couriers to crowdworkers. This form of contract and wage not only is functional in the sense of providing flexibility for employers and pushing costs for downtime, insurance, and work equipment onto the workers themselves, but also serves as a technique to organize the work process in the absence of the physical factory and its supervisors. The piece wage system, as well as the digital possibilities of measuring and controlling the product of work, removes the need to personally control and supervise the worker. The particularities of the piece wage as a form of labor intensification as well as a form of displacement of control of the work process are crucial components of crowdwork. The return of piece wages shows that the phenomenon of digital Taylorism is neither completely new nor the mere return of the old. Instead, digital technology allows for radical forms of the flexibilization of work partially enabled by recourse to a wage form that had previously been marginalized in the capitalist mode of production (although it never vanished).

These conditions aiming at the creation of a hyperflexible and scalable workforce contribute to the precarity of crowdwork. In the sector of microwork, these conditions are more often than not worsened by very low wages. An International Labour Organization survey on five major platforms finds that on average, in 2017, a worker earned US$4.43 per hour when only paid work is considered, and if unpaid hours such as the time used to search for tasks are considered, then the average earnings drop to US$3.29 per hour.[23] A very volatile order situation on many platforms further contributes to the precarity of crowdwork.

The digital platform allows for a sophisticated division of labor, structuring, and organizing the labor process and controlling the worker. As a result, even if crowdwork is a form of home-based work, digital technology allows the real subsumption of labor outside the disciplinary spaces of factories. The labor process is completely organized and structured by the platform, it takes place on a large scale, and it is part of a sophisticated division of labor, hence highly socialized. A special quality of the platform as factory, however, is its ability to enclose a wide range of workers without homogenizing them spatially and subjectively. This touches on a further central argument: unlike the Taylorist factory, digital platform labor does not produce a homogeneous subject; there is no digital mass worker in the Fordist sense. On the contrary, contemporary digital Taylorism allows for the inclusion of a very heterogeneous workforce in very diverse situations, social constellations, and locations. This does not mean that there are no patterns within the workforces of different platforms. However, the heterogeneity of the workers organized by the platforms in a system of tight (if invisible) cooperation and division of labor between workers is stunning compared with the subjective, spatial and organizational synchronicity a traditional Taylorist factory requires.

While the real subsumption of labor, to describe the control of the details of the production process in the famous terms of Marx, that is normally associated with Taylorist factories is generally understood as homogenizing not only the labor process but also the workers and the society surrounding the factory, the platform as a digital factory is more flexible. The platform as an infrastructure of production has little need to homogenize its workers—on the contrary, an important particularity of the platform is that it can be accessed by (almost) anyone, (almost) anywhere, at any time. Accordingly, the technologies of subsumption characteristic of digital Taylorism have the inverse effect of traditional Taylorism: rather than homogenize workers, they facilitate the multiplication of labor. The next section, which presents insights into the demographic and spatial composition of crowdwork, illustrates this.

WORKERS OF THE PLATFORM

Daniel, the student crowdworker based in Berlin, is part of a diverse and globally distributed workforce. On many platforms, tens of thousands of digital workers are active at the same time, sitting in front of their computers solving tasks. While Daniel is working at tasks in between lectures in coffee shops near his university or at night in his small flat in Berlin-Wedding, his coworkers on the multiple platforms he frequents include a wide variety of people in very different locations and situations: Ranging from Indian software engineers working full-time to provide for their families to North American pensioners earning extra income to stretch their pensions, and from Palestinian refugees in Lebanese camps searching for any opportunity to earn money to Spanish single mothers combining crowdwork with care work, the overarching characteristic of this digital workforce is its heterogeneity.

In the digital factories called crowdwork platforms, these people from all over the world are synchronized into one workflow. It is precisely the standardization and algorithmic management in combination with the possibility of logging onto the platform from any place with a stable internet connection that allows the inclusion of such a broad range of workers. Workers can access platforms from their homes, internet cafés, and even their mobile phones. This spatiotemporal flexibility makes new workers and new temporal units accessible to wage labor as such. Without crowdwork, there would be very few possibilities for Daniel to earn money in the thirty minutes between his lectures. This opens up new pools of labor to capital. This development, in turn, contributes to a shifting global division of labor, as well as renewed forms of gendered exploitation, and is ultimately part and parcel of a further flexibilization of the labor market.

Once again, the concept of the multiplication of labor developed by Sandro Mezzadra and Brett Neilson proves a fruitful starting point for the analysis of these dynamics. First, crowdwork is an example of the potential of digital technologies and infrastructures to upset traditional labor geographies and bring very different economic spaces and situations together in real time. Crowdwork, as well as digital technology more generally, is part of an ongoing heterogenization of global space constituting fragmented, overlapping, and unstable cartographies and questioning stable categories such as North/South or center/periphery. Second, the concept hints at the literal multiplication of labor in the sense that many people need to work more than one job to make ends meet. Crowdwork is a primary example of work that can be added to other jobs, paid and unpaid, to survive financially. Often, this includes a further blurring of times of labor and free time. Hereby, crowdwork is a radical example of the flexibilization of labor and the trend toward unstable and multiple labor arrangements. Finally, the term hints at the very heterogeneity of the workers of the platforms.

In the platforms' workforce, we can thus observe a variety of dimensions of the multiplication of labor, from the very literal dimension of people working several jobs to make a living, to the complex and heterogeneous division of labor emerging across the global digital working class. On many platforms, we find people struggling to find employment in regular labor markets, often due to discrimination, social anxiety, or chronic diseases that tie people to their homes. Other workers have a hard time generating enough income with one job. In these cases, digital platform labor often becomes the second or even third job. Crowdwork is often an attractive option to these workers, as it can be fit into any schedule and can be performed in the evening, at night, or on weekends. Some even find time to do crowdwork while on their regular job, as one worker from upstate New York reports. He works in a call center of the regional health care system and has enough time to work on a platform between calls. "That money pays for heating oil for my family in the winter and funds vacations in the summer."[24]

Especially in the Global North, crowdwork has become an important means for many people to add badly needed additional income. Another worker on the Mechanical Turk platform describes that her "husband earns enough with his job to pay the basics, but this income literally puts food on our table. It pays for school outings, Christmas presents, birthday gifts and vacations."[25] As a migrant to the US, she struggles to get job interviews, which is the main reason she has turned to platform work to make ends meet.

These examples show how crowdwork platforms fit into the economic and social dynamics of a global economy hit by multiple crises. It also shows how crowdwork platforms can access temporal units that were hitherto not accessible to capital. In the following, I touch on two important dimensions of this development and two crucial components in the composition of crowdwork labor: women shouldering care responsibilities who now can work on crowdwork platforms while performing domestic labor, and the extension of mobile internet infrastructure in the Global South. Knowledge of this development has fostered several experiments seeking to tap into this labor resource via crowdwork, such as selling internet or mobile phone credit in exchange for microlabor.

CROWDWORK AND CARE WORK

iMerit, a platform that offers digital labor specifically for the development of AI, some time ago posted a description of its platform by a news outlet onto its website to advertise its services:

> There's a dirty little secret about artificial intelligence: It's powered by hundreds of thousands of real people. From makeup artists in Venezuela to women in conservative parts of India, people around the world are doing the digital equivalent of needlework—drawing boxes around cars in street photos, tagging images, and transcribing snatches of speech that computers can't quite make out.[26]

Mira Wallis and I have argued elsewhere that both the reference to women as workers and the comparison of the digital labor behind AI to needlework are not as coincidental as they might seem.[27] And indeed, while the workforce on the platforms is very heterogeneous and often male by the majority, one group that can be found on almost all platforms and among all countries are women who combine crowdwork with unpaid reproductive labor.[28]

Crowdwork is typically done from personal computers in private homes and can be done at any time a worker might be free to spend some time with the tasks available. This spatial and temporal flexibility is well suited to people performing unpaid care and housework—still predominantly women. A twenty-nine-year-old worker from Missouri whose husband's work takes him away from home most of the time describes the importance of the additional income earned on Mechanical Turk as follows: "I am required to be at home as much as possible to be with my children and make sure they are being taken care of. Mechanical Turking has been incredible for my family. I have been able to help pay toward medical bills, and our rising electric bills."[29]

Many women on crowdwork platforms explain that they had to leave their regular jobs to take care of chronically sick or elderly family members and resort to crowdwork to make up for lost income. "I'm retired and am trying to make my retirement dollars stretch as long as possible," explains another worker who was forced to retire from her regular job against her will. "Working full time again isn't an option as my mom is having significant medical issues and I'm her primary caregiver."[30] Such stories are especially common among workers from the US, for example, where many people struggle with the cost of external care services in the context of a health system that gives little security to people with low income.

Young mothers are also found among the workers trying to combine private care work with paid labor on digital platforms. One mother describes how she finds it very hard to earn enough to complement her husband's income and provide for her children: "I find myself spending 8–10 hours a day (whenever I have a spare minute between doing household chores, and taking care of children) to sometimes earn $10.00 a day."[31] Like her, many squeeze in half an hour or an hour of crowdwork in between care labor or domestic tasks to add income for themselves or their families. Many female workers with care responsibilities stress the fact that digital labor via crowdworking sites is their only option to earn money while also caring for children, partners, or relatives. Thus, one important dimension of home-based labor on digital platforms lies in the gendered division of (reproductive) labor and the crisis of social reproduction as it plays out in different contexts globally, from the US to India, as well as in the context of, for example, austerity measures in Italy or Spain.

Digital labor from home done by people with care responsibilities is not only done via classical crowdwork platforms. Several websites are dedicated to "work at home mums" channeling digital labor—whether crowdwork or other forms—to this group of home workers. The North American site wham.com, for example, features articles, job listings, and a forum with more than 245,000 members and over three billion posts, where a variety of online and telephone jobs that can be performed from home while caring for children are shared and discussed by forum members.

Digital technology and infrastructure open up new possibilities for the outsourcing of digital labor into worker's private homes. Home-based labor as such, however, is not new but has a long history. Work done in private homes is a model known, for example, from women's sewing work in the nineteenth century. In *Capital*, Marx cites the examples of lace-making and straw-plaiting in England, done predominantly from private homes and almost exclusively by women and children. The "domestic industry," writes Marx, has become an "external department of the factory." Besides the industrial workers in the factories, "capital also sets another army in motion, by means of invisible threads: the outworkers in the domestic industries."[32] In his description of this system, Marx describes piece wages as the basis for the modern domestic labor, something that is true for historical as well as contemporary examples of home-based work in textile and many other industries with predominantly female workers. These forms of a gendered division of labor and their social and spatial organization are important precursors of today's gig economy and need to be situated in its genealogy.

Interestingly, some of the strategies to devaluate paid homework also exhibit historical continuities. The home as workplace permits the combination of reproductive and care labor, and feeds into the myth of the "bored housewife" who sews—or microtasks, as it were—for fun, and thus does not require a proper wage. Discussing wages and the reason why people work on his platform, the CEO of a German company argued with me that "they do it for fun. Many sit at home in front of their TVs, watch DSDS [a popular television show] and categorize some pictures in passing."[33]

The invisible female labor incorporated into software architectures also reflects computing's gendered history, as well as the relationship between software and the gendered division of labor. A century ago, the word *computer* referred to human workers who executed the commands of scientists on makeshift vacuum tube computers or calculated ballistics tables in military institutions.[34] The majority of these workers were women, and although some of them were trained scientists, the gendered division of labor tended to categorize their work as clerical data entry, while public recognition for the development of computing was reaped by male scientists and engineers, even though female workers became increasingly significant in computing departments during World War II.[35] In her *Programmed Visions*, Wendy Chun writes about that period during in which many of these young women were referred to as computers: "Not only were women available for work during that era, they also were considered to be better, more conscientious computers, presumably because they were better at repetitious, clerical tasks."[36] While some of these women would go on to work as scientists, the majority of human computer jobs were made redundant throughout the sector's further development. These female human computers' position exhibits many similarities with that of today's crowdworkers: just as the labor of the human computers was hidden behind both male scientists and the machines themselves, today's crowdworkers are hidden behind platforms and the magic of AI.

"THE NEXT FIVE BILLION"

"I was a full-time worker with [a] multi-national corporation, India, where I was handling recruitments for my state, before I had my baby who is 2.5 years old now, post which . . . I love working from home for sites like clickworker.com."[37] This worker on the clickworker.com platform also turned to crowdwork because care responsibilities tied her to her house. Unlike the other workers cited previously, however, she is part of the growing digital working class of the Global South. These workers form another important segment of crowdwork, which represents the majority on many platforms.

Many crowdwork platforms have opened their infrastructure to almost anyone with an internet connection, while others concentrate on specific national workforces. Today several platforms specifically target the populations of the Global South in search of low labor costs. The outsourcing of digital labor to the Global South is nothing new as such. For a long time, it was mainly channeled by big firms handling business process outsourcing through national branches or foreign partnerships—with the comparatively cheap digital labor found in India and the Philippines as main hubs.[38] Crowdwork platforms have now begun substituting some of these flows in a more decentralized way, thereby diversifying the Global South's digital workforce in geographic and social terms. India and the Philippines are still among the most common locations of workers on crowdworking platforms, but overall, the geography has become less centralized and online labor platforms constitute arguably the first truly "planetary" labor market.[39]

"We're running our house hold using Mechanical Turk, paying

mortgage, paying for food, supporting family and you know needs are endless and in this if we lost our accounts, that we spent almost half decade to made it and burned our blood by working all day night long."[40] Between these lines of this quote by another Indian worker, one can sense the fear of having the account banned. Amazon has restricted new accounts for Indian workers, which has led to a boom in the trading and sharing of existing accounts and prompted Amazon to begin banning accounts that allegedly provide false information. The criteria by which accounts are banned seem to be quite random, and the fear of having an account banned is among Indian workers' biggest worries.

The turn toward workers located in the Global South is dependent on infrastructure, as access to computers with a (somewhat reliable) internet connection remains limited, albeit growing. The global proliferation of mobile phones, smartphones, and mobile internet infrastructure is thus opening up direct access to an even broader and geographically diverse global workforce: the two to three billion owners of mobile phones in the Global South, most of whom have no regular access to stationary computers with an internet connection. A pioneering project regarding the inclusion of further workers into the global market for digital labor is the company txteagle. Launched in 2009, this platform set its sights on the rural poor, who could solve small tasks in exchange for air time. This was even possible via text message, allowing people without smartphones to participate. The number of cell phone subscriptions in Kenya increased over two hundred-fold from 2000 to 2012, a fact that made it a perfect starting point for this early venture in mobile crowdworking in a country of the Global South.[41] Nokia, for example, used the twenty thousand Kenyan txteagle workers to translate their mobile phone menu into local languages. Txteagle couched its advertising in claims to offer Africa's rural poor opportunities on the global labor market but were also quick to note that these workers were 20 percent cheaper than their Indian counterparts.[42] Today, the company has changed its name to Jana and reaches more than thirty million people, who receive free air time in exchange for watching advertisements and participating in small tasks.

With this, Jana is in line with several similar infrastructural experiments such as Facebook's Internet.org/Free Basics, which seeks to connect the global poor to the internet without making them pay with cash but, rather, with their attention, their consumer choices, and their labor power. This initiative consists of a partnership between Facebook and other companies such as Samsung, Nokia, and Ericsson to connect the "next five billion," as Mark Zuckerberg referred to the people mostly located in the Global South not yet connected to the internet and not on Facebook.[43] The initiative promoted technical solutions and new business models to allow smartphone users in the Global South to access the internet. It seems clear that the rising interest shared by all important information technology companies in extending internet coverage to the Global South is motivated not only by a desire to develop a new pool of customers but also of potential workers. As smartphones increasingly come to embody new forms of payment infrastructures, the possibilities for microwork appear to be huge.

In the context of infrastructural development, Kenya has been inspired by the Indian and Filipino model and is striving for a "Silicon Savannah," a strategy that includes the development of crowdwork.[44] Countries like Kenya have turned to crowdwork as a development strategy and a way to generate an influx of money. Many hopes are tied to crowdwork as a global labor market that includes and empowers workers from the Global South.[45] However, crowdwork platforms also set into motion fierce competition between workers on a global scale, often triggering a race to the bottom, so to speak. Furthermore, while many crowdwork platforms today are global labor markets theoretically open to anyone with an internet connection, access to these platforms is unequally structured by questions of nationality, infrastructure, currency exchange, skills, and discrimination, and workers in the Global South earn on average less than their counterparts from the Global North.[46] Hence, the notion of a seamless global labor market fails to correspond to the complex and fragmented geographies of digital labor, transformed once more by the proliferation of crowdwork platforms. Instead of a borderless global labor market, crowdworking platforms are digital factories producing a complex geography of labor connected to physical and political spaces (e.g., national legal frameworks) in multiple ways. The internet's special quality as a space of production that is connected to physical and political territories but also exceeds these spaces in particular ways is an important part of the heterogenization of global space. In many respects, crowdwork is a primary example of this emerging economic geography.

HIDDEN LABOR

Automation in its various forms, AI and machine learning being only one of many, does indeed replace human workers in the present and has the potential to extinguish even more jobs in the future. The jobs that are automated away, however, have a curious tendency to reappear, although looking often very different and taking place at new locations, done by new workforces. The infrastructures and technologies of automation themselves are produced and sustained to no little extent by human labor. A globally distributed workforce organized through thousands of crowdwork platforms is a premium example of this tendency. The importance of these forms of human labor to the development and support of AI is considerable if often little recognized. For crowdworkers, the current huge investments in AI mean a rising demand for data labor rather than its automation. Of course, this could change in the future, as much of the work done on crowdwork platforms is indeed at many current frontiers of software development. However, these developments can create new needs for living labor, as we can vividly observe it in the present. Automation in the context of globalized capitalism is much less the linear process that many predictions suggest, rather, it is a turbulent, uneven, and crisis-ridden process where sometimes labor is automated at one point only to reappear at another, often recomposed in geographic and social terms and hidden behind new infrastructures of code and concrete.

The labor of crowdworkers is indeed hidden in various ways. Mostly, it takes place outside public spaces, often in private homes. It is geographically distributed and less visible than most other forms of labor. It also takes place outside the reach of many forms of labor regulation and traditional forms of labor conflict. Furthermore, it is hidden behind the magic of algorithms. Much of the work done on digital platforms is masked as software. Much of the labor that crowdworkers do is thought to be already automated. In fact, crowdwork becomes particularly important where software alone is unable to find solutions, as is often the case with visual, contextual, cultural problems. Crowdworking platforms allow corporations to outsource these problems to a globally distributed workforce, while their algorithmic architecture allows for tight and automated control of the labor process, resulting in a hyperflexible and scalable on-demand workforce that can be integrated into complex software architectures.

Crowdwork platforms, just like other gig economy platforms, often claim to be mere intermediaries between employers and employees, somewhat resembling online labor markets. Upon closer examination of the platforms and the way most structure the labor process, however, it becomes evident they are anything but a neutral intermediary between the supply and demand of labor. As infrastructures of production, they assume many of the sociospatial functions of the traditional factory. Labor on microtasking platforms is characterized by a high degree of standardization and decomposition, as well as digital surveillance and measurement of labor. As with any factory, however, the platform as a digital factory engenders questions about not only technology but also the legal and social arrangements in which it is embedded and which it produces. The legal construct of the independent contractor and piece wages not only is a means of creating a flexible workforce but also provides crucial tools for organizing and disciplining living labor. Combining these elements facilitates a specific form of the real subsumption of labor under capital outside the disciplinary spaces of the physical factory or office. This form of organizing labor in combination with growing connectivity opens up new pools of labor to capital. Workers can access platforms from internet cafés, their homes, and even their mobile phones, making new workers and new temporal units accessible to the wage-labor form as such. This in turn has produced a shifting global division of labor, as well as renewed forms of gendered exploitation, and is ultimately part and parcel of a further flexibilization of the labor market.

Dispersed across the globe, isolated from and in competition with one another and hardly protected by labor legislation, crowdworkers' possibilities for collective action appear quite limited. Nonetheless, various forms of resistance have emerged, often starting in online forums where crowdworkers meet, socialize, and support one another. Another pathway is tactical technological interventions such as the Turkopticon, a browser plug-in for workers to exchange assessments of requesters and their tasks, allowing a low-level form of a digital strike.[47] From letter-writing campaigns to collectively written pamphlets on minimum requirements for crowdwork, digital workers have managed to make themselves heard and formulate political demands. These protests show that crowdworkers are indeed able to communicate and act collectively, the difficult conditions limiting their action notwithstanding.

5

THE HIDDEN FACTORY

Social Media

"It began with the dream of a connected world. . . ." These are the opening words of the much-acclaimed documentary *The Great Hack*, which premiered at the 2019 Sundance Film Festival.[1] The documentary, produced by the streaming platform Netflix, chronicles the scandal surrounding the data firm Cambridge Analytica influencing elections, such as the 2016 Brexit referendum or Donald Trump's victory in the US presidential elections in the same year. The documentary scandalizes the illegal harvesting of user data and the bombardment of users with questionable election advertisements that Cambridge Analytica carried out, and which was enabled by the complicity of today's most important social media platform: Facebook.

At the high of its power, Cambridge Analytica claimed to have five thousand data points on every voter. The firm's data sets, profiling software, and targeted advertisement could win elections—this was the not totally implausible claim of the British company. Clearly, Donald Trump's election campaign believed in the potential of targeted advertising: At the height of the race, the campaign spent a daily amount of $1 million for advertising on Facebook. While the extent to which the work of Cambridge Analytica influenced the outcome of the elections in question remains a topic of heated discussions, the importance of social media platforms as opinion factories has become apparent. As Cambridge Analytica became a symbol for voter manipulation and the spread of fake news, the most important platform for its campaigns also came under increasing scrutiny. In the wake of the scandal, the role of Facebook itself became the object of much critique. While its enabling of the practices undertaken by Cambridge Analytica and its handling of the scandal provoked outrage, the critique soon extended to other broader problems of privacy and the spread of fake news and hate speech. Since 2016, images of Mark Zuckerberg pressured by members of the US Congress in various legal hearings have become an almost regular news item.

Indeed, the scandal around the double win for right-wing politics in 2016 marks a turning point in the image of Facebook. The rise of fake news and the concurrent concerns for democratic procedures challenged the image of social media, and Facebook in particular, on a hitherto unprecedented scale. Facebook's innocent claim to make the world a better place by connecting people seems increasingly hypocritical to many, and the platform is at the center of broad concerns and debates over democracy in the age of fake news. To this day, Facebook has not been able to recover from 2016, a year that was as disastrous for Facebook's image as it was for global progressive politics. With its damning critique of user surveillance and data harvesting by Facebook, however, the Netflix documentary also feeds into a line of critique that has been around for years, even before 2016. In many countries, fears around privacy had been the biggest issue for the company for a long time. With Cambridge Analytica, the explosive political force of a practice that lies at the very base of Facebook's business model became obvious: the platform collects user data and turns the attention of its users into a commodity that is sold to advertisers.

At the very beginning of the documentary, this business model is briefly mentioned. Illustrated by shots of demonstrations and racist chants, the film's narrator, David Carroll, a professor for digital media based in New York, asks: "How did the dream of a connected world tear us apart?" Shortly thereafter, he provides an answer as to why this might be the case: "We are now the commodity." In the documentary, this question of personal data being sold is discussed mostly as a problem for democratic procedures. On another level, however, it becomes clear that the harvesting of personal data to sell targeted advertising to companies lies at the core of Facebook's political economy.

Critical analyses of digital media have denounced this unprecedented level of collecting personal data for many years. Furthermore, debate abounds around the question of value and free labor among theorists of digital labor. While Facebook provides the platform, virtually all of the content that makes the site attractive is generated by users, and it is their data and attention that the company sells to the advertisers. This has led several theorists to argue that user activity on social media should be considered exploited labor.[2]

Both this debate and the critical analysis of user surveillance and privacy breaches are debates of crucial importance. However, a set of questions is missing from these debates. The Netflix documentary—just as much of the aforementioned contemporary critique of social media—gives the impression that this data is produced without effort, stored without place, sorted without problems. This is not the case.

This chapter addresses a question that stays mostly in the background of debates around digital surveillance and free labor: that of infrastructure. It delves into in the infrastructure of social media and discusses social media as infrastructure. By taking the algorithmic, material, and human infrastructures of social media (in other words, code, data centers, and the living labor incorporated in platforms) into account, it draws attention to the technological, material, and labor-intensive dimensions of the political economy of Facebook and other social media platforms. Such an approach illuminates a multiplicity of other sites of social media labor: from security guards and technicians at data centers near the Arctic Circle, software programmers and testers in Silicon Valley, to content moderators in Germany, India, and the Philippines.

THE POLITICAL ECONOMY OF PLATFORM ADVERTISING

On a basic level, the business of Facebook and Google is not very complicated: both companies make the overwhelming majority of their revenues through advertising. These two platforms also dominate the global market for online advertising. All estimations on global ad spending, a market of well over $300 billion, see Google as the biggest ad seller. The company accounts for roughly a third of all sales, followed by Facebook (taking in roughly 20 percent of the ad spending) and Alibaba, while Amazon comes in fourth.[3] Income through advertising makes up over 80 percent of Alphabet's (Google's mother company) revenue and over 95 percent of Facebook's. Even though Alphabet is diversifying its business, advertising is at the core of its business model, making it one of the most valuable companies globally. In the case of Facebook, also among the most valuable companies by market capitalization, the case is even more clear: the platform's huge revenues and profits are virtually all based on the sale of advertising space on its social media outlets.

The major distinction between online and traditional advertising is that Facebook and Google can offer precise targeting. Advertisers can target people using basic factors such as age, gender, and location, as well as more refined metrics such as interests and behaviors, due to the data Facebook collects on its users. Facebook is able to gather this data through the user's interactions with the platform and by following the user's online activity outside the platform. This allows Facebook to collect and store a huge amount of data and to construct precise profiles of individual users. Bundled into groups according to established criteria, the attention of those users is then sold to advertisers.

With over 2.5 billion users visiting Facebook at least monthly, it is the biggest social media platform globally. In addition to facebook.com, the company has acquired several other outlets, such as the messenger service WhatsApp and the social network Instagram, both with well over one billion monthly active users. While there are desktop versions for all of these, mobile apps are now the most important form of use and account for the overwhelming majority of advertising income. Looking at the different brands, websites, and mobile applications, David Nieborg and Anne Helmond propose an understanding of the company as "data infrastructure" that hosts a variety of "platform instances."[4] With this conceptualization, they direct attention both to the ecosystem of brands and applications and to the attempt of Facebook to become a crucial infrastructure of everyday life. Facebook and Alphabet's services Google Search and YouTube are not only the world's most visited web pages; they are becoming ever more crucial infrastructures of daily life. This is no by-product of successful platforms; rather, it is the very core of their strategy and political economy. This development implies a form of critical analysis that starts from the platform's infrastructure and moves from there to understand its strategy to become the social infrastructure of the modern online world.

A crucial part of a platform's infrastructure is its software code. The code of a search engine or social media platform fulfills infrastructural tasks in order to create, organize, maintain, and enclose digital activities in a way that harnesses attention in a highly profitable manner. Similar to how Keller Easterling describes the task of infrastructure, a platform's design allows for certain things while making others impossible: "It's not the declared content but rather the content manager dictating the rules of the game."[5] Any platform is designed to stimulate certain forms of expression while preventing others and provides a set of protocols governing the users interacting with the platform and with one another. Therefore, a platform like Facebook strives to connect with other sites of the online and offline world as much as possible, seeking to become the site or host of a growing number of social interactions. At the same time, it accrues a massive amount of data about its users, which encompasses their connections to other users, cities, products, political movements, food, and so on. It aggregates and fragments this data to sell quantified and metrical packages of attention to advertisers.

ALGORITHMIC ARCHITECTURES: LOGIC, CONTROL, LABOR

Platforms like Facebook and Google are complex algorithmic architectures; this section focuses its short analysis on only a small segment thereof. Beginning with Google's famous PageRank algorithm, it moves on to discuss Facebook's Open Graph protocol as an attempt to become the algorithmic infrastructure of the social web. PageRank could be described as a component of the informational web, whereby the search engine registers links between sites and constructs a hierarchical index, while Facebook and its Open Graph protocol correspond to the social web, understood as an asset composed of relationships between people and things and hence a form of social indexing—or, in the words of Carolin Gerlitz and Anne Helmond, the distinction between the "link economy" and the "like economy."[6] Of course, these logics are neither historically successive nor mutually exclusive; rather, they operate simultaneously in different places and in ever new combinations throughout the internet.

Google's famous PageRank algorithm developed by founders Sergey Brin and Lawrence Page arguably remains at the heart of Google's power, even today. It is based on a simple principle: PageRank understands the internet as a system of hyperlinked documents. For Google Search, every page is ranked by the number and quality of links leading to the page, operating on the assumption that every time a person creates a link to another website, that person expresses a judgment about the site. In a second step, these links are qualified—a link from an important page is given a higher value than links from a less important page. By crawling the web and amassing a huge number of such judgments, the PageRank algorithm is able to mine the crowd's collective human intelligence and aggregate their opinions concerning a website's relative significance. This form of free labor is not the only instance of human labor hiding behind the algorithmic search engine. Obviously, there is the labor of coding as well as maintaining the algorithmic architecture of the search engine. Many programmers working both in the Mountain View Googleplex and around the world are constantly working at maintaining and improving the search engine to maintain its global dominance. Google continually changes and refines its algorithm to keep up with the changing patterns of internet usage. Here, human labor plays an integral role. As with most software, more human labor power is involved than that of those understood as its programmers. This workforce is not typically understood as programmers or software engineers and are often outsourced and hidden from view. In the case of Google's search engine, for example, another workforce comes in to refine the results: the so-called raters. This is a group of mostly subcontracted digital workers, often working from private homes, who rate the algorithm's results to further refine it through improved matching of results with queries.

Raters are often part of the huge contingent of home-based digital workers. Organized through crowdwork platforms or subcontracted by specialized firms, these people's contribution to Google's search engine is hidden behind the magic of the algorithm crafted by the company's founders. Working from home, raters log on to an online tool provided by Google and begin judging search engine results according to criteria such as "vital," "useful," "relevant," "slightly relevant," "off-topic," or "spam," or identify pornography. This labor is a crucial component of the ongoing cultivation of the search algorithm and thus shows how every algorithm contains slices of past and present human labor. An important company providing this labor used by Google is Lionbridge, based in Waltham, Massachusetts, which outsources most of the search engine rating tasks to digital laborers working from home around the world. "Are you looking for a job that affords you the opportunity to work with one of America's top 100 most trusted Companies, while working from the comfort of your home?" advertises Lionbridge's home page. "The job involves analyzing and providing feedback on text, web pages, images and other types of information for leading search engines, using an online tool. Raters log on to the online tool to select tasks to do on a self-directed schedule."[7]

One such rating worker describes this arrangement: "I schedule my own hours; as long as I get at least 10 but no more than 20, I stay on pretty good terms with them. They are very strict, but allow you to make up hours that you missed."[8] Lionbridge also has tight productivity goals for its workers: "There is a certain number of tasks that I must complete every minute, depending on the task type. If I fall short of those goals, I am put on probation, during which I cannot work. If my quality isn't up to par, they fire me. It's a very controlled work environment."[9] This is another form of digital labor that is hidden from view, as it takes place in private homes and is still highly controlled and disciplined. It also shows how platforms such as Google's search engine cannot be viewed as a simple algorithm but, rather, as a messy combination of material infrastructures, software, and human labor that is constantly changing.

Although the functioning of Google Search is of course much more complicated than presented here and subject to continuous and ongoing development and diversification, the PageRank algorithm and its simple basic principle are the foundation of Google's power. Matteo Pasquinelli argues that, through PageRank, for the first time "the apparently flat data ocean of the internet was shaped by Google in dynamic hierarchies according to the visibility and importance of each website."[10] Google achieved this feat by combining different forms of labor: the free labor of the people linking to websites as well as the labor of coders or raters writing, maintaining, and refining the algorithm.

Facebook took a step in 2010 to expand its reach beyond its immediate network and follow users through the web, attempting to map metadata in a spirit both similar to and different from Google's PageRank system. At Facebook's f8 conference for developers and entrepreneurs, Bret Taylor, Director of Facebook's Platform Products, in an announcement that arguably constituted the most important development in the platform's history, told the audience that "the Web is moving to a model based on the connections between people and all the things they care about" and presented the Open Graph, an innovation designed to make Facebook the key infrastructure for this form of the social web.[11] Open Graph is a protocol, an application programming interface (API), which uses social plug-ins to allow any page to be labeled with metadata and connected with the Facebook platform. Facebook tools like the famous "Like"-button could then easily be integrated into any object on any website. The incentive for website operators to integrate Facebook into their website through Open Graph is the promise of raised visibility and increased traffic. Facebook's incentive is that it can follow the user through the internet and thereby multiply the possibilities of collecting data on user behavior and preferences.

With Open Graph, Facebook was able to widen the map of its users and their connections on the platform beyond its own platform to include millions of web pages that are now connected to the platform. The index consists of website metadata that turns the sites and objects into nodes on Facebook's graph, as well as users' interactions with these sites and objects. The latter can be considered the more significant component, as they develop an architecture that works by the logics of personalization and peer group matching, creating a map of the social web based on "the connections between people and all the things they care about," as described by Taylor at the f8 Summit.[12] It goes without saying that becoming the owner of such a map is a very attractive prospect for a company like Facebook. The Open Graph is an infrastructure for the collection and ordering of data that enriches Facebook's knowledge of its users' behavior outside the platform, which in turn is highly lucrative information for advertisers. Beyond its own page, Facebook has successfully installed itself as an infrastructural grammar of "liking" and "sharing," inscribing itself into the semantics of wide swathes of the internet.

The "Like" button was initially the Open Graph's essential function and remains important to this day, but the protocol has since developed to allow more functions. The "Share" function allows users to share content with their friends, which has proven particularly crucial in the context of news articles. Today, almost every news site or blog features a plug-in allowing readers to share articles with their friends and contacts. This has profoundly changed the nature of news and information. Many internet users no longer visit specific news sites, instead of navigating the news by reading articles shared by friends in their news feeds on various social networks. In this sense, Facebook's platform has also become a metanews page, where every user finds a personalized collection of articles curated by their peer group and Facebook's algorithm. In this way, Facebook (and, in another form, Google as well) not only plays an important role in deciding what news stories users see but also sets standards and creates dependencies that change the production of news.[13] Facebook has become much more than a social media platform, and Facebook's Open Protocol system is thus only one example of the (highly successful) way Facebook seeks to inscribe itself in the infrastructural grammar of the social web.

PageRank and Open Graph are only small parts of the algorithmic architectures operated by Google and Facebook, an architecture that is constantly changing. An analysis of such software (especially if it goes deeper than the perfunctory observations noted previously) contributes immensely to understanding the political economies of these corporations. These algorithmic systems guide and channel users in ways that map their behavior while functioning as architectures of enclosure and extraction designed to valorize small increments of human attention and labor. Google and Facebook are more than a mere search engine and social media platform. Rather, they strive to become the critical infrastructure for their users in as many sectors and areas of networked life as possible. This is, however, not just a question of algorithmic infrastructures, but also one of hardware.

THE MATERIALITY OF THE CLOUD

Facebook's home page has more than one billion daily users and is the third-most visited website on the internet after Google Search and YouTube. One simple click on the home page "requires accessing hundreds of servers, processing tens of thousands of individual pieces of data, and delivering the information selected in less than one second," as Facebook describes the processes behind its business.[14] The number of photos and videos alone stored on Facebook's servers exceeds one hundred petabytes (one hundred quadrillion bytes). To manage these amounts of data, Facebook needs a huge and manifold material infrastructure. Alongside its headquarters in Menlo Park, California, and seventy national and international offices such as the European headquarters in Dublin, Central London, and Hyderabad, India, the other important class of buildings held by Facebook are its data centers.

With the rise of cloud computing, the data center—or rather a certain variant of the data center—has been transformed into a central pillar of the internet's infrastructure. The in-house server, whether integrated into the computer or in the server room of an office, is increasingly being replaced by huge data centers handling computing and storage for a variety of networked and spatially distributed devices. In contrast to the associations its name evokes, the cloud is not immaterial, nor does it replace hardware; rather, it represents its spatiotechnological reconfiguration on a global scale. This entails not only a centralization of infrastructure into huge server farms, but also job losses in the IT departments of many companies—some of which reappear in the massive data centers where the labor of data infrastructures is outsourced, centralized, and streamlined in an "industrial mode of production, processing, distribution, and storage" as Vincent Mosco writes in his study on contemporary data centers, *To the Cloud: Big Data in a Turbulent World*.[15] The dynamically evolving global geography of data centers and their related infrastructure are a crucial expression of the materiality of the internet. They represent a site where questions of the digital economy and its data infrastructure, as well as environmental issues, the digital transformation of sovereignty, and new forms of the transformation and outsourcing of labor are negotiated today.[16] The transition from mainframe computers and in-house servers to a planetary network of data centers and a system of storage capacity and software on demand is connected to the rise of social media.[17]

In its first years of existence, Facebook relied entirely on leased data center space. In 2006, two years after its founding, growing traffic on the social network almost led to a server meltdown. At that time, the company rented a forty-by-sixty-foot space in Santa Clara, California, to house its servers. One day, the servers could not cope with the rising internet traffic, overheated, and were at risk of melting down. Facebook's chief engineer sent out employees to buy every fan they could find in the area to cool down the server and avoid offline time.[18] Since then, both the number of users and the data infrastructure have changed significantly. Despite the incident, it took Facebook four more years to begin construction of its own data center in Prineville, Oregon, signaling the start of a strategic shift, as Facebook now strove to manage all of its computing and storing processes in its own data centers. Completed in 2011, the Prineville site has been expanded continuously and today encompasses more than three million square feet; it is still Facebook's largest data center complex. The roughly 160 people who work at the complex consist of managers, engineers, cleaners, and security.

Facebook has built more gigantic data centers in the US and abroad since 2011. One of the major centers, which opened in 2014 and has grown steadily since, is located in Altoona, Iowa. Another huge complex is located in Forest City, North Carolina, and includes two buildings spanning more than three hundred thousand square feet each, one dedicated to storing "cold data," data which is not accessed regularly by users. North Carolina has become a major hub for data centers in the US. Apple's data center, valued at over $1 billion, in Maiden, North Carolina, a rural town of some three thousand inhabitants, is among the largest in the world. Thirty miles northwest of Maiden in Lenoir, North Carolina, Google has invested $1.2 billion in a similar complex. North Carolina grants generous tax breaks and infrastructure upgrades, as well as offers cheap power and comparatively cheap labor. Cheap land, energy prices and reliability, climate, and the availability of cheap and qualified labor are crucial factors in deciding to build data centers at a given location.

Facebook opened its first major European data center in 2013, located close to the Arctic Circle in Lulea, Sweden. The cold climate helps to keep servers cool, a major issue for large-scale data centers. Facebook is not the only market actor moving its data centers to Northern Europe, where operations can be up to 60 percent cheaper than in central Europe. Google, for example, opened a data center in Hamina on the Gulf of Finland in 2011, where the cold seawater is used in the cooling system. Microsoft and Google have radicalized this approach by investing in research in underwater data centers, and Microsoft has filed a patent for an underwater data center designed as an artificial reef.[19] In addition to cooling advantages, the ocean floor is a somewhat more stable environment, with fewer disturbances to be expected from factors such as storms, fires, or politics. In Facebook's Lulea data center, the tremendous need for energy (another major concern for data centers) is met with hydroelectric power. The Lule River produces about 13.6 million megawatt-hours of hydroelectric power, covering 10 percent of Sweden's energy demand and making energy comparatively cheap and reliable. The reliability of Lulea's energy supply also convinced Facebook to reduce its backup generators by 70 percent compared with its North American facilities. All data centers seek to offer continuous service with no downtime which is another major factor in their environmental footprint: not only the backup diesel generators and chemical batteries but also the center's high use of power overall has attracted widespread criticism.

When the song "Despacito" by Louis Fonsi broke a record and reached over five billion clicks on YouTube in 2018, its energy consumption equaled that of forty thousand US homes in a year, according to estimations.[20] Data centers alone account for more energy consumption than countries like Iran, and the carbon emissions of the information and communications technology ecosystem as a whole is on a par with the carbon footprint of the aviation industry.[21] As early as 2010, the environmental organization Greenpeace released a report on cloud computing and its contribution to climate change, explicitly targeting and criticizing Facebook for its first data center in Prineville designed to run primarily on coal-fired power stations.[22] Greenpeace launched a successful campaign against Facebook called "Unfriend Coal," which generated considerable attention and pressure on Facebook. The decision to build a "green data center" in Lulea should be viewed in this context. Along with Facebook, many other data center providers came under attack for the environmental impact of their infrastructure, prompting corporations to initiate programs to "green their clouds." The energy consumption of data centers and other computing infrastructures illustrates the very materiality of the cloud, which appears literally quite dark indeed if one happens to look at the diesel generators in the data centers at the right time.

Lulea was also the first data center to run exclusively on hardware developed in Facebook's Open Compute Project. Each data center houses tens of thousands of networked servers connected to the outside world through fiber-optic cables. Facebook has not only invested in its own data centers but also diverted massive funds into technological developments such as cooling systems and server technologies, not least through the said Open Compute project. The volume of data not only rises with the number of users but also with the development of technology such as the ever-higher resolution of photos and videos taken with a common smartphone and the advent of virtual reality technology, which produces amounts of data that dwarf the average photo or video. At the time of this writing, Facebook was expanding most of its existing centers and building several new ones, including a $250 million investment in Los Lunas, New Mexico, a facility projected to cost over $1 billion in Fort Worth, Texas, and a second large European center in Clonee, Ireland. In 2017, Facebook announced its third European data center would be constructed near Odense, Denmark, with more to follow. Like other major digital corporations, Facebook is now among the most important owners of data center infrastructure in the world. Accordingly, although some major corporations such as Equinix or Digital Realty continue to concentrate on data centers, the prominent digital corporations such as Facebook, Microsoft, and Apple have now become major players in the field of physical internet infrastructure as well, with Amazon Web Services as the market leader in cloud computing and server hosting.

In addition to its own data centers, Facebook also leases server capacity from other providers. Besides various sites in the US, Facebook has been leasing capacity in Singapore and is building its own data center there. The Singapore data center, with an estimated cost of one billion dollars, will be an eleven-story, 170,000-square-meter high-rise center as the island faces growing land constraints. Singapore has been a hot spot for data centers for some time, for example, for Western cloud companies looking to serve Asian markets. Singapore is favored for its business-friendly government as well as major undersea cables laid along telegraph lines dating back to British colonialism.[23] Both midsized and large Asian firms like Alibaba operate their own data centers there, and it has also become an important location for big Western players like Microsoft and Google. The locations of data centers are chosen not only for their infrastructural advantages but also for political reasons. Google's Asian data centers in Singapore and Taiwan in particular are attempts to stay close to but outside of the legislative reach of the Chinese state. Following a two-year stand-off with the Chinese government over censorship issues, Google redirected its Chinese search engine to Hong Kong in 2010 and soon announced plans for a data center there as well. In 2013, two years after the ground-breaking ceremony, the plans were abandoned and investments re-rerouted to Singapore and Taiwan. This illustrates both the complex relation of cloud computing to the transformation of sovereignty and the material embeddedness and complex political geography of its infrastructure such as submarine fiber-optic cables and data centers.[24] These new politico-infrastructural geographies are not to be understood as mere conflicts between states and transnational corporations but, rather, as the transformation of sovereignty itself through infrastructure, in a multilayered, conflictual, and dynamic process. Keller Easterling has coined the term "extrastatecraft" to account for the multiple, overlapping, and nested forms of sovereignty that are produced by large-scale infrastructures and their relation to national states.

UNDER THE SEA AND INTO THE FACTORY

Speaking to a *New York Times* reporter, Jayne Stowell, a "Strategic Negotiator, Global Infrastructure at Google" commented: "People think that data is in the cloud, but it's not. It's in the ocean."[25] She is referring to submarine cables that provide another crucial component of digital infrastructure. Even if the rise of wireless devices suggests otherwise, cables are a crucial internet infrastructure, especially fiber-optic undersea cables that speed up intercontinental network traffic.[26] Stowell's main job is to oversee the construction of Google's undersea cable projects. In fact, Facebook, Google, Microsoft, and Amazon have become major players in this business. Facebook and Google teamed up in 2016 to build a submarine transpacific cable from Los Angeles to Hong Kong. The 12,800-kilometer fiber-optic cable will have a capacity of 120 terabytes per second, the highest of all transpacific cables.[27] Earlier that same year, Facebook announced a partnership with Microsoft to build a submarine cable connecting Virginia Beach, Virginia, with Bilbao, Spain. This is part of a wider trend whereby the bandwidth needs of cloud-powered firms like Google and Facebook are so great that they must build their own submarine cables rather than rely on purchased capacity from other carriers (normally consortia of private and public actors).

Among the projects Jayne Stowell is overseeing for Google is "Curie," a 10,500-kilometer-long undersea cable connecting the United States with Chile, the location of Google's most important data center in Latin America. This cable was built and is owned exclusively by Google, as are two other new cables connecting the US and France and Portugal and South Africa, respectively. The sole ownership of such expensive and complex infrastructure is a novelty even for the biggest tech companies and illustrates the size and economic force of Google's operations. The fiber-optic cable for Curie was built in a factory in Newington, New Hampshire, and has about the width of a garden hose but is encapsulated in protective layers of steel and copper. In the nearby Piscataqua River, a specialized ship took the cable aboard in a labor-intensive, week-long loading process and traveled to California, the site of the new cable's start. Working in twelve-hour shifts, the ship's crew then started the slow process of laying the cable on the seafloor.[28]

Google's and Facebook's moves toward their own submarine cables represent yet another expression of the immense infrastructural power the companies have accumulated, demonstrating how they have become major players in the political and economic development of internet infrastructure. The security guards, cleaners, and engineers in the data centers, the workers in the cable factory, and the workers spending months aboard the ship laying the cable also hint at the vast amounts of labor incorporated in the physical infrastructure of global platforms. These workers are part of a large and diverse workforce involved in producing the material infrastructures for social media. Surrounding the comparatively small core workforces employed by platforms like Facebook and Google are concentric circles of primarily outsourced workforces that take part in the production of the various infrastructures that allow Facebook or Google to function. Another of these infrastructures is the devices that allow users to connect to social media platforms such as laptops, tablets, and smartphones, as, for example, Apple's famous devices.

IPHONE CITY

Most people today probably know that a large portion of Apple products is assembled by a company called Foxconn. Foxconn became the infamous global face of contract manufacturing under harsh working conditions in 2010 when the plight of workers in Chinese Foxconn factories was highlighted by a series of worker suicides, most of which were committed by rural migrants in the new industrial zones.[29] Many of the individual Apple product parts are produced in other locations before being shipped to the big Foxconn plants, where the final assembly occurs. One crucial part of this supply chain is the semiconductor factories where the chips for the various devices are produced. The most important production site for semiconductor technology is Taiwan. The huge factories of the Taiwan Semiconductor Manufacturing Company are also the source for most of the chips in Apple's products. The whole process is tightly controlled by Apple.

China is the production center for Foxconn, even though rising labor costs and the trade war between the US and China have led to a spatial diversification, such that some of the iPhone's assembly has been moved to India, among other places. Foxconn still employs hundreds of thousands of workers in China, for example in Zhengzhou, also called "iPhone City," situated in the region of Henan, one of China's poorest provinces. Depending on the season and the order situation, up to three hundred thousand workers are employed at the sprawling complex and most of them live in nearby dormitories. Their main tasks are final assembly, testing, and packaging of iPhones, which means most workers engaging in one step of these tasks, such as polishing screens, do these steps about four hundred times per day.[30]

Another cluster is in the Pearl River Delta, one of the earliest, biggest, and most important sites for the outsourcing of production of IT products globally. Although Foxconn has been relocating some elements of production away from this area in response to workers' protests and rising wages, it remains a major center of their Chinese operations. Foxconn's biggest facility in the Delta is the Longhua Science and Technology Park near Shenzhen, also known as "Foxconn City," a complex employing hundreds of thousands of workers. In Pun Ngai's book *Migrant Labour in China*, a worker describes her job at the assembly line at the Longhua factory: "I take a motherboard from the line, scan the logo, put it in an anti-static-electricity bag, stick on a label, and place it on the line. Each of these tasks takes two seconds. Every ten seconds, I finish five tasks."[31]

The Pearl River Delta is also an expression of a model of production and infrastructural development based on zones and corridors. These zones, among them the Pearl River Delta, were also the first sites of market economy experimentation in China.[32] Shenzhen, one of the Delta's biggest cities, was declared the first Chinese special economic zone (SEZ) in 1980, initially attracting Taiwanese and Hong Kong firms before becoming a center of contract manufacturing for Western corporations as well. The SEZ was repeatedly expanded until it encompassed all of the city's districts in 2010. The city has also undergone explosive growth since the 1980s, making it increasingly difficult to establish the exact number of residents, almost half of whom are migrant workers. Together with Shenzhen, the surrounding Pearl River Delta has grown to become one of the world's most important export-oriented production zones. The entire Pearl River Delta is on its way to becoming a megacity and the largest urban area in the world in terms of both size and population.[33]

Before the rise of the Pearl River Delta as a center of IT manufacturing, the United States—Silicon Valley, in particular—had been the most important site. Today, most IT manufacturing has left Silicon Valley; the few factories that remain are characterized by a sharp distinction between highly qualified and well-paid scientists and engineers, and manufacturing workers who themselves are fragmented along subcontracting chains ending in homework, as some subcontractors dispatch work to be performed at home. Work in both Silicon Valley's IT manufacturing sector and the growing service sector catering to the technology companies and start-ups is characterized by precarious conditions, often very low wages, and a high proportion of female and migrant workers.[34] Accordingly, Nick Dyer-Witheford argues that Silicon Valley was built on migrant labor, before extending its class dynamics onto a world scale.[35]

From Silicon Valley, IT manufacturing has moved to several sites such as the Pearl River Delta, Taiwan, Mexico, Malaysia, and eastern Europe, in a process that had been occurring for decades but was accelerated by the 2001 recession. Like the manufacturing workforce in Silicon Valley in the 1970s, global IT manufacturing workers today are predominantly female and to a large extent migrants.[36] Global IT manufacturing across many locations is characterized by a labor regime Stefanie Hürtgen and her colleagues describe as "flexible neo-Taylorism": a contemporary form of Taylorist production, albeit without Fordist forms of regulation, security and orientation toward mass consumption of the workers—a "Taylorism after Fordism."[37] This export-oriented model is based on flexible contracts, subcontracted labor, and low wages (even by local standards) enabled by standardized production technologies, allowing for a high degree of labor fluctuation. Foxconn's plant in the Czech Republic is another example of these tendencies; there, the percentage of subcontracted workers, often migrants recruited through temporary work agencies, exceeds 50 percent at times.[38] These workers produce various devices on which users can access social media platforms like Facebook.

On the devices produced in these sites, the users of Facebook and other social media will be able to see content uploaded by other people such as holiday pictures, view status updates about the local football team, or receive an invitation to a neighbor's birthday party. But even though all content is created by users, not all of the content users upload makes it onto the screens of other users.

CONTENT MODERATION: "THE STUFF YOU SEE IS BEYOND WHAT I COULD IMAGINE"

Every minute millions of comments, status updates, pictures, and videos are posted on a social media platform like Facebook. While normal users see an overwhelming majority of mundane and unexceptional posts, the pictures, videos, and comments Roberto sees on Facebook are anything but mundane: From morning to evening, his screen is flooded with more expressions and depictions of violence, racism, and hate he could ever think of. Roberto is a content moderator. He works for a company that is contracted by Facebook to keep its network as clean as possible. While these attempts to offer users a clean network vary in their success, Roberto experiences the dark underside of social networks every day: "The stuff you see is beyond what I could imagine people could do before I started this job."[39]

Five days a week, he enters an office building in Berlin, goes to his desk, opens his digital moderation tool, and enters the dark side of social media. In his queue, an ever-greater number of "tickets" (content flagged as inappropriate that he has to moderate) is piling up. He is expected to solve hundreds of tickets daily: "You are a machine, clicking all the time."[40] He and around six hundred of his colleagues, the majority of them migrants like him, work in the Berlin offices of Arvato, an important player in digital content moderation. They are part of a global workforce consisting of hundreds of thousands of digital workers trying to guard social media networks, video platforms, dating apps, messengers, newspaper comment sections, and many more digital spaces that allow user-generated content in the way the companies running these platforms want them to be moderated. These workers are a crucial and often hidden component in the infrastructure that makes social media possible. Although mostly hidden from view of Facebook users and the public, Roberto and his colleagues understand their importance in a time when Facebook has become an ever more important network and, in tandem, an important platform for hate campaigns and the spread of various kinds of violent content: "We always say doctors and policemen see the same amount of blood and violence as we do—but they are not paid minimum wages."[41]

Those digital workers are an integral part of the platform's functioning; more specifically, they are closely integrated with its algorithmic infrastructure. As in the case of crowdwork, living labor is integrated into algorithmic architectures to fill gaps in computing where human cognition is required for decisions computers cannot make. Content moderation belongs to the most important segments of human labor hidden behind the digital interface of the platform. It is an extremely labor-intensive, politically sensitive, and economically vital aspect of digital social media. The sheer scale of content and the complexity of content moderation is also an unprecedented challenge to social media platforms.

GOOD AND BAD CONTENT

In recent years, the problem of fake news and hate campaigns and the question of a platform's responsibility for the content posted by users has moved to the center of public debates. In fact, it might be today's single biggest challenge to all big social platforms, from Facebook to YouTube, TikTok to Instagram. Arvato, the company where Roberto works, was hired by Facebook in a reaction to public critique in Germany focusing on the role of Facebook in right-wing hate campaigns. A local moderation team coordinated by a subsidiary of the high-profile German Bertelsmann company was an attempt to appease public critique as well as lawmakers in the process of crafting a new law holding social media platforms responsible for the content uploaded onto their sites. All over the globe, platforms have faced similar critique and legal pressure, making their content moderation efforts a key issue for senior management at all tech companies.

Of all the content uploaded onto social media platforms, a huge percentage may not be liked by other users. On most platforms, users can "flag" this content and thereby start the process of content moderation. Users flag posts for multiple reasons. Maybe they simply don't like the content of a post. Posts may contain violence, racism, nudity, drugs, or a multitude of other content considered offensive by legal or cultural standards. A social media company like Facebook, however, has its own reasons to erase such material from the platform. First, it wants to maintain the platform as space where as many users as possible want to spend their time, and second, many legal and political reasons compel Facebook to take down some of the uploaded content. To avoid being banned from national territories, Facebook tries to appease governments threatening to block the social network. A manual leaked in 2012, for example, shows Facebook's effort to please the Turkish state: "all attacks on Ataturk" were to be banned, along with maps of Kurdistan, Kurdistan Worker's Party (PKK) symbols, and pictures showing Abdullah Öcalan (with the addendum "IGNORE if clearly against PKK and/or Ocalan").[42]

While so-called community standards roughly outline what kinds of content are prohibited on Facebook, the corporation remains secretive about its precise content restrictions as well as how the platform is monitored and by whom. In recent years Facebook has published some documents in an attempt to make its moderation process more transparent. However, information on the precise rules dictating what content moderators allow on the platform is still very hard to obtain, and the publicly available rules and standards are fairly generic. Furthermore, workers report that rules change on a weekly or even daily basis and are quite complex.

Content moderation in general is a very complex endeavor, as interpretations of violence, pornography, humor, and terrorism, to name a few key topics, vary widely across legal, cultural, and political settings. While Facebook has historically been hesitant to take responsibility for content hosted on its platform, considerable pressure has since led the company to invest significant efforts into its content moderation system.

ALGORITHMS PERPLEXED BY CULTURE

Facebook's platform is organized by an increasingly smart algorithmic architecture managing user behavior on the site. Facebook and other platforms invest considerable effort into the development of automated and learning content moderation software. However, there are many fields where algorithmic intelligence regularly fails—for example, cultural norms and practices and their highly contextual nature. Today, Facebook's software can detect nudity in pictures and videos very efficiently (even though these efforts are again hindered by further cultural rules; e.g., Facebook allows male nipples but forbids female nipples). In any case, most content that is removed based on nudity and sexual activity is now preselected by software with a precision of over 90 percent. In the case of hate speech or bullying, on the other hand, the software has a hard time making sense of these highly contextual situations and is incorrect in the majority of these cases.[43] Throughout all categories, Facebook's machine-learning software is only proposing content to be deleted; a human content moderator makes the actual decision at this point.

The goal clearly is a system that works on autopilot. It is doubtful, however, if this horizon is reachable. While Facebook executives often suggest publicly that AI is the solution to the company's content moderation problems, software engineers and specialists are less optimistic. The fact that Facebook now has over fifteen thousand digital workers doing content moderation shows that, at the moment, human labor is an increasing rather than decreasing component of platform moderation infrastructure, despite all efforts of automation. Human cognition remains central to content moderation, and a future in which decisions about what can stay on a platform and what must go are completely automated remains far off.

The labor of reviewing suspicious content is part of a sophisticated global division of labor. Most social media companies have in-house departments dedicated to content moderation. In-house departments normally develop criteria, oversee subcontractors, and handle difficult cases such as politically sensitive issues or cases where law enforcement agencies are involved (threats, violence, child pornography, etc.). These workforces include specialists and lawyers who devise content moderation policies as well as specialists who react to potential high-profile cases and security concerns such as imminent attacks. If Roberto, who works for a subcontractor in Berlin, becomes aware of a post indicating an imminent violent attack, for example, he "escalates" the case, and specialists in Facebook's European headquarters in Dublin will assess the post and potentially contact national law enforcement agencies.

The firm Roberto works for stands for another layer in Facebook's content moderation system. Despite comparably high wages, Facebook has started content moderation operations in countries such as the US and Germany, partly as a result of political pressure in recent years but also due to the insight that a lot of cultural knowledge beyond language skills is necessary for content moderation. In most cases, these operations consist of call-center-like offices, sometimes with several hundred employees run by subcontracting firms but tightly controlled by Facebook, Google, or the other platforms. Another important group, probably the largest segment of content moderation labor, is outsourced to countries like India and the Philippines where local workers also do content moderation for other countries. Finally, another segment of content moderation is outsourced directly into private homes—for example, via crowdworking platforms. The result is a complex and global division of labor with workers in drastically different situations and locations handling the task of cleaning social media platforms.

BERLIN, AUSTIN, DUBLIN: OUTSOURCED MIGRANT LABOR

Roberto works in a call-center operation run by a subcontractor. Responding to growing criticism by German media and politicians regarding the company's inability to react to problems like hate speech and racism, Facebook announced the hiring of Arvato, a German firm, and a subsidiary of the Bertelsmann Group, which employs more than seventy-two thousand people worldwide and offers a variety of services including cloud computing, logistics, and finance, as well as a variety of customer relations, call center, and content management services. Arvato describes itself as one of the many companies providing a wide range of services while remaining largely unknown to the public. "You may not know it," the company wrote on its website, "but Arvato is behind a great number of the products and services you use. On average, every consumer in Germany comes into contact with us eight times a day."[44] In 2019, Arvato merged with the Moroccan Saham Group to form the company Majorel, a leading specialist in content moderation with fifty thousand employees in twenty-eight countries. Majorel now also runs the center in Berlin where Roberto is employed.

In 2016, Arvato hired approximately six hundred workers, distributed in different language-based teams including German, Arabic, English, Turkish, Swedish, Italian, and Spanish. The majority of the workforce is of migrant background and young. Many have university qualifications or other professional degrees that are often not accepted in Germany. "Berlin is perfect for the company. Many migrants from all over the world need work and language is the biggest issue for most," reflects Roberto, who came to Germany a few years before and tried various freelance jobs before starting at the Arvato offices in Berlin.[45] Arvato drew on the growing pool of qualified young workers migrating to Germany, whether fleeing the war in Syria or the effects of the Euro crisis and harsh austerity measures in Southern Europe. Berlin's cultural appeal, comparatively low rents, and the hope for employment have triggered a huge influx of these diverse strata of migrants. However, most struggle to find permanent jobs, and many are unemployed or work in the service sector, often in very precarious and informal conditions, despite often having university degrees and other professional qualifications. With their formal and informal qualifications as well as their difficulties on the German job market, young migrants from various countries constitute an almost perfect labor pool for Arvato and the job requirements of content moderation.

The employment of migrant workers for content moderation is not exclusive to Berlin. In many ways they make for the perfect workforce from the perspectives of the content moderation firms: these workers bring important language and cultural skills while having fewer alternatives in local job markets. Speaking to a reporter, a worker at a subcontracting firm doing content moderation for YouTube in Austin, Texas, explains: "When we migrated to the USA, our college degrees were not recognized, so we just started doing anything. We needed to start working and making money."[46] Many of his colleagues in YouTube's largest content moderation facility in the US are also recent migrants who have worked as delivery drivers or security guards before becoming content moderators hoping to achieve conditions similar to full-time Google employees. Instead, they are part of the workforce Google refers to as TVCs (temporary, vendor, and contractual employees), a contingent that constitutes over 50 percent of Google's workforce. Accordingly, the conditions at the subcontractor in Austin differ greatly from those of Google's full-time employees in Mountain View. The great pressure to perform and the exposure to too much violent content has caused mental problems and symptoms of stress disorders among many workers in Austin.[47]

When Roberto started his job at Arvato in Berlin, he was forced to sign a confidentiality agreement and was forbidden from disclosing that he works for Facebook. This is among the reasons why he and many of his colleagues are very careful when they speak about the conditions at the northwest Berlin content moderation facility. Most of the contracts are temporary, and some workers are hired through agencies, so many are afraid of losing their jobs. The fear of being fired notwithstanding, a few months after Arvato started working for Facebook in Berlin, discontent began to spread among the workers. They complained about pressure to perform, a lack of breaks, and particularly about the highly traumatizing content they are forced to look at for hours on end. After Arvato failed to respond to complaints properly, some workers began speaking to local activists and considered talking to the press about their working conditions, although most were afraid of losing their temporary contracts. Some workers report being required to review more than 1,500 cases ("tickets") per day, leaving them with fifteen seconds per ticket on average.[48]

Finally, the national newspaper *Süddeutsche Zeitung* broke a major story about the conditions at Arvato in December 2016. The story largely focused on the traumatizing images the workers were forced to see and quoted former and current Arvato workers about their experiences: "The pictures were worse, much worse than in training. . . . Violence, sometimes disfigured corpses. People are leaving the room frequently. Run out. Cry."[49] Another worker reported high pressure to reach daily quotas: "You have to reach your daily goal, otherwise there will be trouble with your supervisors. The pressure was enormous."[50] The article stirred up a great deal of attention, and Facebook and Arvato publicly claimed that they provided sufficient emotional and medical support for workers to cope with the distressing experiences—something the workers themselves deny.

After the scandal, Facebook pressured the subcontractor to provide better working conditions. The company lowered the quotas and hired staff to support the workers coping with the traumatizing content. While the lowering of the high quota proved to be a relief for workers, their impressions of the psychological support staff show the extent of distrust between workers and management: "There is a team for psychological support, but they are only social workers and we don't trust them because they work for the company."[51] While the contracts of many workers are not renewed to avoid having to hire them permanently, others leave because of the traumatizing content. Roberto, who explains that he tries to conceptualize violent images he sees "as a movie," is among the few workers who have stayed longer at the job.

In fact, most workers leave within the first two years, either by choice or because they are let go. Sarah Roberts, who has conducted pioneering and extensive research on content moderators, reached similar conclusions. Many of the workers she interviewed in the US were humanities graduates with liberal arts degrees looking for their first job, typically working on temporary contracts and often for various subcontractors. They left the job after a few months or were phased out of their contract after two years, but they still struggled with the violent images they were forced to view.[52]

A worker who quit after three months describes her experiences at Arvato Berlin in a report: "Texts, pictures, videos keep on flowing. There is no possibility to know beforehand what will pop up on the screen."[53] Much of the content is banal, flagged by users because they dislike it for some reason. However, she was shocked by the amount of violence she faced while doing content moderation for Facebook: "The content is very violent. I had been exposed previously to real world violence as I worked in peace building and humanitarian aid in conflict torn contexts. Yet I was far from imagining . . . that violence could be so predominant on social media."[54] She quit her job when she developed a sense of "hyper vigilance" in looking out for her family, started dreaming about her job, and felt that her perception of violent shootings started to shift. "The terrible Las Vegas Shooting suddenly seemed entirely normal to me."[55]

INDUSTRIALIZED DECISION MAKING

The ability to cope with the traumatizing material is hindered by the high pressure to perform present at most content moderation facilities. With the increasing scale and labor intensity of content moderation, this sector becomes a more and more rationalized operation. "The search for uniformity and standardization, together with the strict productivity metrics, leaves not much space for human judgment and intuition," the former Arvato worker explains. "The intellectually challenging task tends to become an automated action, almost a reaction."[56] To organize what Facebook's former head of content policy, Dave Willner, calls "industrialized decision making,"[57] Facebook introduced a workplace software to maximize efficiency that was rolled out in different countries and at contracting firms in 2018. The software, called Single Review Tool (SRT), structures the workday of a content moderator. The moderator sees the different tickets (moderation tasks) lined up in different queues sorted by topics such as "violent extremism" or "nudity," with different priorities. Depending on the queue, the time allocated to moderate can be very short, starting at a few seconds per ticket. Quotas are constantly changing in different sections and locations but are always a means of pushing workers and also a reason for termination. Sources from Facebook's second major German content moderation facility in Essen, for example, speak of a culture of hire-and-fire, where workers who fail to reach certain metrics are quickly sorted out.[58] The SRT software documents all metrics, and the workers must log every break. If the workers are inactive for several minutes, the tool shows their status automatically as "unavailable" to management.

A random sample of finished moderation tasks is controlled by quality assurance, constituted of both workers in the same office and higher-ranked and directly employed workers at Facebook offices in other locations. Quality assurance generates an accuracy score for every worker that is visible in the SRT tool. If workers fall under a certain quality score (98 percent in most places), they face pressure from management. Chris Gray, a former Facebook moderator who worked for a subcontractor in Dublin, Ireland, and is now suing Facebook with claims of "psychological trauma" due to poor working conditions, was fired because of his quality scores. He was asked to achieve an accuracy of over 98 percent in his quality scores, which caused considerable stress if his score dipped beyond the threshold: "So if you come in, and it's Tuesday or Wednesday, and you've got five mistakes . . . all you can think about is how to get the point back."[59] The firing of workers based on low-performance metrics is a normal procedure at many different contracting firms in different countries.

The reports from different sites and contractors in Europe and the US resemble one another and constitute the picture of a working regime characterized by high pressure, quotas, and increasing standardization while the workers are confronted with highly traumatizing content. At the same time, the majority of the work is done by outsourced contractors, often migrants, who tend to hold only short-term contracts, a tool that serves both to discipline the workforce and to react to changing demand. These content moderation facilities in Berlin, Austin, or Dublin hence show many of the characteristics of the specific combination of digital Taylorism with flexible labor relations analyzed across the chapters of this book.

THE GLOBAL GEOGRAPHY OF SOCIAL MEDIA'S DARK SIDE

"Nowadays everybody has access to the internet. And if it is not controlled well, it becomes a porn factory." This is the reflection of a seasoned content moderator featured in the documentary *The Moderators*.[60] This short film, directed by Ciaran Cassidy and Adrian Chen, profiles the induction week of a group of newly hired content moderators in an office somewhere in India. Like the seasoned moderator organizing the training course, all the other executives featured in the documentary stress the social importance of their future work to the newly hired moderators.

Working for an Indian operation that offers to source content moderation for customers from abroad, the new recruits are part of another important segment of the global workforce in content moderation. In Berlin, Austin, and Dublin, the digital factories performing content moderation for social media are comparatively new. In Manila or Hyderabad, in contrast, these digital factories have been in existence for many more years. In the system of content moderation by big platforms, outsourcing to the Philippines, India, and other countries has been and still is a crucial component. A large share of content moderation, also for customers from the Global North, is performed in the Global South. Outsourcing content moderation not only brings in additional profits through comparatively low labor costs but also allows Western corporations to distance themselves from possible consequences such as lawsuits by traumatized workers.

In the documentary, the new workers are profiled throughout the five-day training program, after which they will move to their cubicles and start moderating themselves. For most of the training, they can be seen in front of a big screen, while their trainers explain the rules of their new job. Most of the workers are happy about their new jobs and hope for careers in India's burgeoning digital economy. They also know very little about content moderation. As they will be working for a Western dating platform, among other clients, they are warned by their trainers that their religious worldview might be offended. Throughout the days of training and a flood of exemplary cases, the newcomers seem to oscillate between being humored and shocked. "It's strange, definitely, those pics I saw. That was strange to me. I never thought I'll work on those things, nudity and all," remarks one of the new workers to his colleague as they reflect on their week of training.[61] While nudity seems to bemuse most of the workers, numerous cases of violence shown in brutal detail on the big screen leave the workers shocked while the trainers explain with care and details which rules apply for depictions of torture or violence against children.

Several clocks showing different time zones on the walls of the open-plan office symbolize the international character of this operation. Teams are at work all day and all night, guaranteeing around-the-clock services to customers in all time zones. As content moderation for a dating app is one of their tasks, their jobs at this office are especially sensitive: "People go there to find their soulmate, so they are very vulnerable there," remarks one trainer, emphasizing the importance of their work. Up to 70 percent of the new accounts at the platform are created by fraudsters trying to exploit customers in various ways. The trainer knows very well that the work his office is doing is fundamental to the existence of social platforms, dating or otherwise, as we know them. "The dating industry would not be flourishing at is it is today, online, if moderators aren't there."[62]

"ALMOST IDENTICAL TO WORKING WITH AMERICANS OR AUSTRALIANS"

Besides India, the second important location in the global division of content moderation labor remains the Philippines. As one of the most important locations of outsourced IT labor, the Philippines is home to not only content moderation companies, but thousands of local and transnational companies offering all kinds of digital services. The Philippines today is a central location for the so-called business process outsourcing (BPO) in the IT sector. The sector employs over a million workers across the country and reaches revenues in the range of $30 billion. The general service sector has become the economy's most important, contributing the majority of the country's gross domestic product and employing more than half of its workforce.[63]

Content moderation is among the most important of such services offered to international customers. Together with India, the Philippines have become a crucial location for the outsourcing of content moderation. Facebook, for example, has spread its content moderation teams across twenty countries, including Roberto's Berlin office and contractors in countries such as Latvia and Kenya. However, according to the company, India and the Philippines are the most important outsourcing locations. Besides Facebook, all major platforms, including YouTube and Twitter, outsource some of their content moderation through contractors such as Accenture or Cognizant to the Philippines. The Philippines has a large pool of qualified workers, many with university degrees who have very good English language skills, as English is one of the country's two official national languages. This comparatively cheap and qualified workforce is the foundation of the BPO "success story," in the process of which the

Philippines has overtaken India as the call center capital of the world and become a magnet for the outsourcing of a wide array of IT labor, of which content moderation is only part.[64]

Beyond language skills, call center labor and content moderation demand cultural competencies. Here, the colonial history and postcolonial present of the country become a special factor in the global competition for outsourced content moderation and call-center work. Microsourcing, one of the major content moderation providers to Western social media, praises the advantages of the local digital workers as comparatively cheap but well educated, loyal, honest, and hardworking before mentioning that as "a former US colony with a 90% Christian population, the Philippines has a very 'westernized' culture." Working with Filipinos, according to Microsource, "feels almost identical to working with Americans or Australians."[65]

The Philippines' trajectory to becoming a content moderation hot spot for Western corporations includes, among other things, the history of colonization first by Spain and later the US, postcolonial American influence exemplified by language and the education system, its entry into globalized markets particularly through semiconductors and other electronic products, and an extraordinarily mobile labor force (represented, for example, by Filipino sailors and care workers). The Philippines has a close cultural relationship to the US, as all schools teach English, often with an American accent (something very important for call-center work). Around half a million Filipinos graduate from university each year, most of whom are very familiar with American culture and, having often worked or studied in the US, constitute a rich labor pool for IT labor such as content moderation, which requires a high degree of cultural knowledge. Additionally, the overwhelming majority of the Philippines' population is of Catholic faith as a result of Spanish colonization—another important factor concerning cultural values, especially with liminal cases such as pornography. This shows that the labor of outsourcing, or at least some parts of it, is not something that can be done by everybody. Some content moderation is relatively easy—categorizing pictures, for example. Many tasks, however, require high-level competencies. Content moderators must work according to a complex set of rules and foreign legal standards as well as cultural norms and tastes that are not their own, sometimes in a foreign language.[66] This process is yet another instantiation of virtual migration: while workers remain in their home countries, they work in the cultural, legal, and sometimes temporal framework of another country and are faced with manifold cultural, social, and legal problems often similar to those of real-world migrants.

The Philippines host many large-scale content moderation providers such as Majorel (formerly Arvato), TaskUs, or Microsourcing, as well as many smaller outfits. The number of workers these firms employ is sometimes estimated to be in the millions. By Philippine standards, many content moderators earn a good wage, as do many other BPO workers in IT and customer services. Most are young university graduates, often women from middle-class or poor backgrounds. While most work in large offices in the Manila region, some are located in the new tech parks used by many IT and BPO outfits in other parts of the country. With comparatively high salaries and desks in air-conditioned offices in the modern IT business parks, a lot of content moderators are located in the spatial and social neighborhood of many other digital workers in the BPO sector.

Content moderation labor, however, deserves a special note: workers regularly confront the darker side of a culture that is not their own, and the emotional aspect of the labor is very hard for many to endure. Scanning up to six thousand pictures or one thousand videos every day, many of which contain brutal violence or pornography, leaves many workers with emotional trauma and disturbances. Speaking to the *Washington Post*, a moderator working for a contractor for Twitter remarks: "At the end of a shift, my mind is so exhausted that I can't even think." The Manila-based worker recounts that he occasionally dreams about being a victim of a suicide bombing or car accidents and concludes: "To do this job, you have to be a strong person and know yourself very well."[67] Employees report depression, sleeping disorders, and even affective or sexual problems, while some younger workers have become paranoid after viewing so much child pornography and find it difficult to leave their children alone with anyone else.[68] It is difficult to predict what the long-standing effects of these forms of labor will be, as no substantial studies have been conducted on the long-term psychological effects of content moderation.

Although the Philippines offer cheap labor power compared with Western standards, content moderation remains a form of qualified labor that must solve more complex problems. Outsourcing to the Philippines is still more expensive than the next layers of content moderation, namely, countries with lower wage strata and outsourcing to workers around the globe via crowdworking platforms. As most social media platforms are highly secretive about their content moderation systems, it is difficult to estimate what percentage of content moderation is performed via crowdworking. A leaked manual from crowdwork platforms, for example, proves that Facebook used crowdwork for content moderation in the past, but if and to what extent this is still the case is hard to say. However, a large number of jobs on different crowdworking platforms that are part of different forms of content moderation suggest that an important share of content moderation labor is still done by crowdworkers scattered across the globe. Many crowdworkers report tasks that are clearly part of some content moderation system; either they are deciding over "live cases" or are used for quality control or to train content moderation algorithms.

BECOMING INFRASTRUCTURE

The labor of content moderators underscores how labor-intensive social media platforms truly are. Hidden from users' view, a heterogeneous and globally distributed group of digital workers continuously scrub social media clean. Almost seamlessly integrated into the algorithmic architectures of the platforms, they form part of the complex infrastructures behind social media networks like Facebook. This labor force includes both highly qualified and well-paid experts in the legal departments of huge corporations and an army of outsourced and precarious workers around the world. The latter group in particular can be described as another instantiation of digital Taylorism — closely integrated into algorithmic systems, their labor is organized by computational systems and fills the gaps of AI. Much of their labor is highly repetitive and even boring, but highly emotionally demanding and conducted under precarious circumstances. Whether performed by Syrian refugees in Berlin, young graduates in

Manila, crowdworkers in Tunis, or work-at-home mothers in rural North America—the labor of content moderation has developed a complex and multifaceted international division and geography of labor. The labor of content moderation is not limited to Facebook or even other social media; a wide range of websites, such as newspapers with comment sections, chatrooms, online games, or dating sites allow interaction and some form of user-generated content and thus require human labor to moderate it. These "digital cleaners" are a crucial yet mostly hidden component of the political economy of the internet as such and of social media in particular.

Companies like Facebook and Google have—compared with their revenue and importance—a small number of direct employees. Around this core workforce, however, are concentric circles of workers who make platforms possible. The labor of coders, raters, hardware engineers, and content moderators tends to be hidden behind soft and hard infrastructures and often blend into these infrastructures. Often subcontracted, outsourced, or even working in private homes, they are nonetheless characterized by tight control and discipline. Their labor is thus a sign of both the spatial explosion of the factory and the continuing importance of its diagrammatic function in many sites across digital capitalism.

This labor is a crucial component of platforms' attempts, particularly those of Facebook and Google, to become crucial infrastructures of everyday life. To inscribe itself in everyday life and to become an irreplaceable infrastructure lies at the heart of the strategy of many platforms. This strategic horizon is clearly prominent in the case of Facebook and Google, but also in other companies such as Amazon. The extent to which Facebook today is "platform-as-infrastructure" is, finally, also a reason why the continuing bad press since the Cambridge Analytica scandal has not led to a decline in its user base.[69] Despite all bad press and public scrutiny, the social network's number of users has continuously increased globally, and it has continued its efforts to interweave itself into the fabric of everyday life.

6

THE PLATFORM AS FACTORY

Conclusion

The Bethlehem Steel Company came to Baltimore, Maryland, in 1916. By this time, the company was already among the most important industrial companies in steel and shipbuilding in the US. It bought steel factories and property at Sparrows Point, a peninsula in the Southeast of Baltimore. Close by, the company bought farmland and started to build a town for its workers named after the town of Dundalk, Ireland. Over the following decades, Dundalk attracted many labor migrants who came to work in the factories and would grow to house over one hundred thousand workers by the 1960s.[1] By this time, Bethlehem Steel was at the peak of its power and one of the world's largest steel producers and shipbuilding companies. By the middle of the twentieth century, the company employed hundreds of thousands of workers, contributed the steel parts of the Golden Gate Bridge, produced one ship per day at the height of World War II, and had been the heart of the small city of Bethlehem for over a century.[2] The original site was over time complemented by other factories across the country, such as the complex in Baltimore, employing over thirty thousand workers at its peak.

Frederick W. Taylor made pivotal contribution to the success of Bethlehem Steel. The father of scientific management joined Bethlehem Steel in 1898 to solve the firm's machine capacity problems. At this time, Taylor was already well known; his rationalization strategies and management techniques had engendered broad interest among factory owners and managers, and his efforts to rationalize production driven by precise studies of workers and labor processes received a great deal of attention. Bethlehem Steel became the setting of one of Taylor's most important studies and also an example in his most important publication. The tale of the "man we will call Schmidt" is a central part of Taylor's *Scientific Management,* in which he describes his experimentations with various forms of disciplining and incentivizing techniques by the example of a handler of pig iron at Bethlehem Steel.[3] It is an important step in the development of Taylor's theory of management based on the systematic gathering of knowledge over the labor process, the monopolization of this knowledge in the hands of management, and tight control over every movement of the worker. The separation of planning and execution, the decomposition and standardization of tasks, and the precise surveillance of workers not only were crucial to rationalizing the workflow but must be viewed as attempts to systematically break worker's resistance.

Scientific management, or Taylorism, is often described as the product of Taylor's work at Bethlehem and other production facilities. Conversely, one might argue that it simply gave a name to a development that was already taking place all over the United States and Europe. Management techniques such as those employed by Taylor could only take place in the context of certain technological developments, an increasing division of labor, and the further socialization of the production process. The extent to which Taylorism is a product of Taylor's questionable genius or to which it is simply a result of broader sociotechnological developments remains open.[4] In any case, the names Taylor and Bethlehem Steel now signify a crucial step in the development of factory manufacturing and the subsumption of labor under capital in a way that moved the factory further into the center of society.

Today, however, Bethlehem Steel is history. The long decline of the company ended in bankruptcy in the early 2000s and serves as one of the most prominent symbols for the demise of industrial production in the US. The five huge blast furnaces at its main factory in Bethlehem, once a symbol of North American industrial manufacturing power, now serve as the backdrop for an entertainment district. Sparrows Point, in southeast Baltimore, has also transformed profoundly and is now a logistics hub. The site of one of Bethlehem's

Dundalk factories is today occupied by a new kind of factory: an Amazon distribution center.

BETHLEHEM TO AMAZON: TAYLORISM, THEN AND NOW

Another Amazon warehouse across the street came to reside at the site of a former General Motors factory, a similarly symbolic postindustrial takeover. The 4,500 workers at the two Amazon Baltimore warehouses earn about half of what their unionized predecessors made. However, this is still above the sectorial average in a city shaken by postindustrial demise and high unemployment. Many of the workers who could only secure part-time jobs at the Amazon distribution centers still must depend on food stamps issued by the city.[5] At the Baltimore distribution center, just as at any other Amazon FC around the world, workers are subject to a tight labor regime. While it has advanced quite a bit compared with Taylor's experiments at Bethlehem Steel, its logic and principles clearly show its heritage.

At Amazon today, the organization, measurement, and disciplining of labor are increasingly dependent on digital technology. Striving toward an integrated view of all relevant processes and the extraction of as much data as possible, these software architectures are designed to accelerate and make more efficient the movement of people and goods through time and space. Software is thus also of crucial importance to the organization of living labor in logistics. Amazon's distribution centers, in Baltimore and elsewhere, are spaces saturated by software. Each worker's productivity is automatically measured and compared to others; those who fail to reach the quotas are let go. Documents that became public in the context of a labor conflict show that in only one year, Amazon fired at least three hundred workers because of their low productivity at the newer Baltimore facility. Amazon's algorithmic system tracks the productivity of each worker and automatically generates warning and terminations in case of substandard productivity, according to documents obtained by *The Verge*.[6] Sometimes, it seems, digital technology allows for a radicalizing of Taylor's concepts. In a certain sense, digital measuring of the labor process fulfills a historical wish of scientific management, whose precise studies of the laboring body have been described as data mining avant la lettre.[7]

Throughout this book, I have used the term *digital Taylorism* to conceptualize related developments appearing throughout the world of labor, including standardization, decomposition, deskilling, automated management and human computation, algorithmic cooperation, digital measurement, and surveillance of labor. These are all elements of an emerging tendency of work found in different variations and combinations across all the sites visited through the chapters. In my understanding, the term digital Taylorism does not describe a simple rebirth or continuity of twentieth-century scientific management but is rather used to conceptualize how digital technology in particular allows for the mobilization, renewal, and recombination of crucial Taylorist principles in novel ways and contexts. One difference of many is, for example, the temporality of management. While Taylor, Frank and Lillian Gilbreth, and others faced a back-and-forth between their studies and improvements in the production process, digital Taylorism's horizon is a system of real-time control, feedback, and correction. In this sense, the growing importance of algorithmic management based on sensors, networked devices, and integrated software architectures can also be interpreted as a form of a cybernetic Taylorism striving for real-time management and correction of problems.[8] Furthermore, networked devices, sensors, and apps have moved Taylor's time and motion studies outside the enclosed spaces of factories and distribution centers and into the urban space of the logistical city. Amazon Flex or UPS drivers, bicycle couriers, and other workers on the crucial last mile are also increasingly managed and overseen by software. Digital Taylorism is hence no longer bound to the disciplinary architecture of the factory. Indeed, today's digital factory itself can take many different forms.

The workers in Amazon distribution centers, the Flex delivery drivers, the crowdworkers on Amazon Mechanical Turk, as well as Chinese gold farmers, Facebook's content moderators, Google's search engine raters, and the book scanners mentioned in this book's introduction—all of these are workers of today's digital factory. The concept of the digital factory is at the core of this book and is aimed to shed light on a specific tendency among the current transformations of labor driven by digital technology. This entails an approach of theorizing contemporary capitalism not by the factory's end, but by its transformation, multiplication, and generalization. The chapters of this book focus on sites where digital technology produces labor regimes characterized less by creativity and autonomy than repetition, decomposition, and control. These forms of labor are shaped by digital technology in ways that are usually not foregrounded in theorizing the contemporary period of digital capitalism and critical theories of immaterial, creative, or cognitive labor. The book is thus an attempt to shift the spotlight onto sites of labor that appear less frequently in such debates to provide a fuller picture of social cooperation and the division and multiplication of labor under digital conditions.

At the moment, digital technology allows for the tight organization, control, and measurement of labor processes outside the enclosed space of the factory; the digital factory takes very different shapes. It might be an Amazon distribution center in Baltimore, a crowdwork platform organizing hundreds of thousands of home-based workers across the globe, or a large gold-farming workshop somewhere in China crammed with computers running twenty-four hours per day. The concept of the digital factory is not meant to deny the differences between these sites. Even if those sites share a surprising number of common features that make the framework of digital Taylorism meaningful, this perspective is not designed to level out the crucial differences between them.

Furthermore, digital technology has not only allowed a peculiar actualization of scientific management: in different places, it has produced very different labor regimes. It is important to underscore that the impact of digital technology on the world of labor is both sweeping and difficult to capture in a single formula, expressed as it is in heterogeneous ways. In approaching this question of the digital transformation of labor, with this book I thus took double precautions: First, I was careful not to view one labor regime as the only or dominant expression of labor in digital capitalism. Instead, I foregrounded the interplay of several markedly different labor regimes as crucial to capitalism in general, and its contemporary variation in particular. Digital Taylorism is an important tendency in the contemporary world of work, but it coexists—and must coexist—with other labor regimes that show different characteristics. Second, I was careful to avoid proclaiming the absolute novelty of contemporary labor regimes, instead seeking out continuities, echoes, and reconfigurations of earlier labor regimes.

THE PLATFORM AS FACTORY

Amazon has woven itself into the fabric of urban life, in Baltimore just as in other cities. It delivers cloud computing for private companies and public institutions. It owns food stores, and its lockers are distributed across the city. Its online platform has changed the game for local businesses who now depend on their ability to sell their products through Amazon's platform and delivery network. This delivery network consists of many different parts; it includes Amazon Prime Air jets parked at Baltimore-Washington International Airport as well as gig workers using their cars to deliver packages through the Amazon Flex program. At the same time, Prime Video produces movies and series designed to further tie the customers to the platform, and Amazon is venturing into health services and home automation. The company has become a factor in many different areas of the contemporary city. This is an expression of the rise of contemporary platform urbanism, where Amazon and other platforms, such as Uber and Airbnb, are striving to become crucial urban infrastructures.[9]

Clearly, Amazon's goal is to become an irreplaceable infrastructure of everyday life. This characteristic unites Amazon with a company like Facebook that works in a very different field. In varying but often similar ways, the strategy of different platforms is directed at becoming infrastructural. This sort of strategy always aims at monopolistic positions and is by nature expansive. Philipp Staab, a German sociologist, argues that the crucial characteristics of digital capitalism are how companies like Amazon aim at not only being a dominant force in the market but becoming the market themselves. Their goal is to create new kinds of "proprietary markets," where the rules of exchange are dictated by the large corporations that own the markets.[10] Amazon's e-commerce platform is a prime example of this. The example of Amazon also shows how much this strategy relies on a range of material things such as distribution centers, server parks, and cargo jets. Although the firm would prefer to present itself as a technology corporation run by the magic of algorithms and a few genius Silicon Valley programmers, platforms like Amazon, Google and Facebook are all dependent on extensive material infrastructures and diverse labor forces, as has become clear throughout the different sites of this investigation.

Large segments of these workforces are employed in flexible and contingent arrangements. A novel characteristic of the digital factory is its ability to combine the tight organization of the labor process with contractual flexibility and forms of contingent labor. In fact, the chapters in this book show in various contexts how the means of algorithmic management and digital control allow the management of flexible workforces. These workforces, in turn, become ever more heterogeneous. Throughout the different instantiations of the digital factory, we can observe the multiple ways in which the standardization of the labor process can drive, participate in, or profit from the heterogeneity of living labor. This multiplication of labor becomes visible across many sites. In Amazon's warehouses, for example, the various technologies of standardization and algorithmic management reduce training times and increase control possibilities, thereby allowing flexible and short-term solutions in the recruiting of labor to satisfy the contingencies of supply chains. Seasonal labor, short-term contracts, and outsourced labor are important components of the labor regime in Amazon's distribution centers.

Digital platforms, however, are engendering these logics in the maybe most radical way. If one follows the packages out of the Amazon distribution center onto the last mile to the customers, the emergence of platform labor becomes very visible. In the context of an ever-expanding on-demand logic, logistics processes on the last mile of delivery are becoming increasingly important, and demands on speed, flexibility, and efficiency are rising. Here, app-based forms of labor are particularly present and make the last mile of logistical operations a frontier of the gig economy. It is no coincidence that platforms like Amazon's delivery program Flex are becoming increasingly significant in the area of urban logistics and mobility. The search for flexible and scalable labor resources has fueled the rise of platforms such as Flex, Uber, and Deliveroo and has increased platform work's presence in urban logistics and beyond.

Today, platform labor is pushing into ever new sectors. From taxi services like Uber to courier and delivery services like Deliveroo, from the portal Helpling, which brokers cleaning services, to platforms for every kind of digital work, like Amazon Mechanical Turk—there is hardly any sphere of the social division of labor and everyday life in which digital platforms do not play a role.[11] App-based work on digital platforms is characterized by algorithmic organization, direction, and supervision of the labor process on one hand, and by flexible contractual forms on the other. In many ways, the paradigmatic digital factory of the contemporary might be a platform.

These platform workers delivering food, cleaning flats, and driving taxis share a great deal with their online counterparts working on crowdwork platforms. They also provide a specific form of on-demand labor that is even more flexible and scalable—"people-as-a-service," in the words of Amazon's Jeff Bezos. This analogy to software-as-a-service is not incidental. The labor of the crowd is often hidden behind algorithmic infrastructures and organized in a way that allows its insertion into complex software architectures, while cooperation between workers is automatically organized by the algorithmic de- and recomposition of tasks. Crowdwork platforms are another variant of the digital factory; their form of digital decomposition, automated labor management, and surveillance allow for the inclusion of deeply heterogeneous workers without the need to spatially and subjectively homogenize them.

These infrastructures open up new labor pools previously difficult or impossible to reach for wage labor and further diversify the workforce of the digital factory. On crowdwork platforms, Indian software engineers work alongside jobless Moroccan youth, North American ex-convicts, German single mothers, struggling Spanish freelancers, and young Filipino graduates. Some hope for a career in IT, many need additional income to supplement other jobs, while a few merely do it for fun. Crowdwork plays a role in the infrastructural indexing of new digital labor resources in the Global South and is implicated in the digital reconfiguration of the gendered division of labor. The possibility of computer-based homework allows people occupied with care and domestic labor, more often than not still predominantly women, to also become digital wage laborers. Here, it becomes very clear that digital technology is part and parcel of a deep transformation that not only changes the labor process or opens up new ways of controlling workers but that is profoundly restructuring the social and global division of labor.

The efficiency of platform labor, however, is not simply a question of digital technology but also of flexible contracts and various forms of piece rates. Easier than payment measured by time, piece wages can substitute a form of direct control over the labor process because pace and intensity of labor are directly coupled with payment. This way, a certain part of the conflict inherent in exploiting the capacities of labor power is displaced onto the individual worker. This is based to some extent on the standardization of tasks, and often on systems of automated management of workers (e.g., based on "reputation," digital measurements of previous clients' satisfaction) as tools to manage labor biographies. Time, made fluid by the digital renaissance of piece wages, is an important component of both the flexibilization and intensification of labor.

While digital technology allows for an extension of such arrangements, it is important to consider the longer trajectories of multiple forms of flexible employment. Piece wages, for example, have a long history, and Taylor experimented with them at the Bethlehem Steel Company in his attempt to raise productivity. Platform labor is not a detached, utterly novel phenomenon; it should instead be viewed in the context of other forms of contingent labor. The rise of the kinds of labor relations typical of the gig economy cannot be understood simply as a result of technological disruption but must be situated in the context of broader and predigital developmental tendencies, particularly the flexibilization and precarization of labor relations and mobility practices across highly diverse contexts. Logistics can serve as an example of this long tradition of flexible, variable, and informal contractual arrangements, and hence place the current discussion around the gig economy into historical perspective. The history of piece wages, day labor, and informal arrangements began long before the advent of digital technology. Thus, it is necessary to contextualize platform work both regarding the present, such as temporary and agency work, as well as from a historical perspective regarding things like female home-based wage labor and day laborers, which are currently undergoing a specifically digital comeback facilitated by platforms. Such forms of contingent labor are historically always gendered and disproportionately performed by migrant and mobile workforces.

It is, then, no coincidence that the workers on today's various platforms of the gig economy are often in their majority migrant workers. Today, migrant and racialized workers are a crucial part of the on-demand workforces of the digital factory. Throughout the gig economy and beyond, the digitally organized labor process combines with mobile labor in various ways, leading to a high level of migrant platform workers in cities all over the globe. Various factors, including flexibility, precarity, pay, and language determine how platform work is in a specific way incorporated into a stratified labor market, tied to diverse migration routes and projects, bridges waiting periods at other jobs, or complements other wage labor or employment forms. The workforce of the digital factory is flexible and highly mobile.

MOBILE LABOR, FLEXIBLE BORDERS, DISTRIBUTED STRUGGLES

The geography of work and its reconfiguration in the context of digital infrastructure has been another important vector along which this book has researched the transformation of labor. The reordering of global space not only by logistical operations and digital infrastructures but by the very mobility of workers themselves is a central factor in the global recomposition of class in the digital age. Today, there are hundreds of millions of people who are categorized as migrants. These mobile populations are of course also workers, and their mobility shapes the past and present of global capitalism.

Looking at the very heterogeneous forms of labor mobility today, from highly paid programmers to undocumented workers in agriculture, from migrant platform workers to Chinese virtual migrants, the importance of the mobility of labor can hardly be underestimated. While whole sectors of the global economy are based on the exploitation of migrant workers, it is the very mobility of labor that continues to be a challenge to accumulation. The "autonomy of migration" is as material as the differentiated and often violent tools aiming at blocking and channeling human mobility.[12] The operations of some of these control devices result in the production of a special status or, rather, a range of special statuses for migrant workers, from undocumented labor to all forms of visa and permits, which in turn produces new forms of accumulation enabled by the drawing precisely on the different status of the heterogeneous contingent of migrant workers. This is part of the productive power and strategic role of borders for the accumulation of capital—a role that is inherently incomplete and always highly contested, first and foremost by migrants themselves, both at the various sites of border struggles but also through their role in labor conflicts across the globe.[13]

Digital infrastructures and their reconfiguration of space are also implicated in this multiplication of the figures of migration. In the contemporary, we tend to find multiple forms of virtual, temporal, partial migration bound up with multiple geographies of "differential inclusion," in the words of Sandro Mezzadra and Brett Neilson, rather than binary demarcations of exclusion and inclusion.[14] A variation of this multiplication of migration is embodied in the figure of the Chinese gold farmer, who might be an internal migrant to an urban area and at the same time a virtual migrant to the Western servers of World of Warcraft. This example shows another way in which digital infrastructure and connected devices reconfigure economic and labor geographies and create new spatial constellations. Digital technology is part of the constant reconstitution and reformulation of (economic) space, creating new connections and proximities, as well as new fragmentations and borders. The shadow economy of gold farming becomes possible both by the option to connect to the game from anywhere and the unequal positions of those connecting to it. The digital space of an online role-playing game opens up new connections, economies, and geographies of labor yet produces no smooth space of global communication. Participation in this economy is deeply related to the local space from which the game is accessed while transforming those spatial constellations at the same time.

Online games such as World of Warcraft are examples of such a simultaneous proliferation of connectivity and borders. In the servers populated by Western players, Chinese or Venezuelan digital workers become virtual migrants whose labor is profoundly racialized—only one of many similarities these virtual migrants share with their offline counterparts. The production of new experiences and geographies of virtual migration is not determined by infrastructural vectors alone. Factors such as education, language, and culture are also important issues regarding potential workers. These very issues have made the Philippines a global center for digital social media content moderation. Its colonial and postcolonial history has produced an English-speaking, Catholic workforce well versed in North American culture, and hence very much qualified to perform content moderation for Western social media. Their labor exhibits an important affective dimension inherent in experiences of virtual migration.

The term *virtual migration* serves as a conceptual provocation and opens up lines of inquiry for migration research, complicating the widespread definition of *migration* as a physical movement of people across borders. How are these new forms of digital mobility to be understood conceptually and legally, and also concerning lived experience and subjectification, or reconfigurations of racism? Digital technologies produce, structure, and form a large portion of all mobility practices today, particularly in the sphere of mobile work, and thus point to the need for a reevaluation of fundamental concepts and theories of migration.[15]

As the digital factory has been characterized by migrant workforces, so have the numerous struggles affecting digital factories around the globe. In the US, a contingent of Somali migrants started the first strikes in Amazon distribution centers on US soil: the strike action at a distribution center close to Minneapolis–Saint Paul airport amid the 2018 Christmas season was most probably the first coordinated strike at an Amazon warehouse in its home country and was coordinated by an organization called the Awood Center with the motto "Building East African worker power." This migrant coalition was the first to organize a strike at Amazon in the United States, and they were also the first group to force management into negotiations and achieve some successes.[16] From the strikes and conflicts in the logistics warehouses in Northern Italy to the protests and actions by Uber drivers and Deliveroo riders in Paris to the content moderators at Arvato Berlin protesting against unbearable working conditions, struggles in the digital factory are often carried by migrant workforces. This tendency is remarkable, as these workforces are often in especially vulnerable conditions, such that not only contingent contracts but often also precarious residency situations render them especially exposed to retaliation by employers.

Generally speaking, the struggles in and against the digital factory look very different and take various forms, from tacit gestures of refusal to full-blown strikes. It is of little surprise that the heterogenization and fragmentation of class results in diverse and fragmented struggles. With the "end" of the big industrial factories, political certainties have vanished along with the clear-cut subject of antagonism. Accordingly, much political and theoretical effort has been invested in the rethinking and reconstitution of such a subject. The multiplication of labor in all its dimensions also poses a challenge for both theorizing and organizing, which oftentimes is still oriented toward the normative model of standard employment. Moreover, unions face increasingly transnational geographies of production, often characterized by unclear legal regulations and jurisdictional responsibilities. It is often difficult for major unions to come to terms with the expansion of various modes of freelancing and develop adequate tools of struggle, although there have been considerable efforts in this direction in recent years. In some cases, small grassroots unions are better able to adapt to new conditions, as they are often less restricted and more innovative in their tools of action, from the logistics warehouse in Northern Italy to the various strikes and protests of courier drivers at Deliveroo, Foodora, and other platforms.

The spatial diffusion of the digital factory—exemplified in the case of crowdwork, a factory run by tens of thousands of workers at the same time who are spatially separated from one another—is in many cases another obvious hindrance to the development of forms of collective struggle. The form of collectivity generated more or less automatically by working in a factory under one roof must be actively created in the case of many digital factories, as the example of many platforms shows. This is not only a problem of workers being located across different continents; more importantly, it is also a question of completely different conditions that workers working together on the same platform, and even on the same tasks, experience. The flexibilization of the contract and the multiple forms of outsourced, temporary, or irregular labor arrangements are clearly another driver of fragmentation in labor populations—and these arrangements serve in many cases the obvious purpose of quelling labor unrest. Amazon's distribution centers are an example of the strategic use of contingent labor, but they are also an example that shows that these difficult conditions can be overcome and that many forms of resistance are possible.

A crucial dimension of successful struggles will be the development of new tools of digital organizing. Arguably, crowdwork platforms, with their dispersed workforce of independent contractors, again exemplify this challenge best. These exceedingly difficult circumstances notwithstanding, the question of how to organize effectively around a crowdwork platform is already a subject of experimentation. Attempts are often based on online forums and other forms of online networking, generating actions from letter campaigns to activist technology such as the Turkopticon plug-in and collective legal conflicts, in addition to forms of platform cooperativism.[17]

Digital technology has become a crucial wager in these struggles. Not only has this wager been conjured up as conflicts have arisen over new forms of surveillance or rationalization enabled by digital technologies: one of the most powerful ways digital technology is employed in labor conflicts today is by evoking the specter of automation. Amazon is not the only company that likes to spread news about progress in automation in times of labor conflicts, utilizing the imagery of a robot waiting to take over the place of a striking or ineffective worker as a powerful mode of disciplining living labor. Amid the campaign for a $15 minimum wage in the US, a huge billboard could be spotted in San Francisco that read: "Meet your minimum wage replacement" and depicted an iPad ready to take the order of a customer. "With a $15 minimum wage, employees will be replaced by less costly, automated alternatives," the billboard argued. In a similar spirit, the *Wall Street Journal* commented in 2017 that the Minimum Wage Act should be called the "Robot Employment Act," arguing that "$15 an hour doesn't help poor youth. It helps Flippy, the new burger-grilling machine."[18]

The Fight for $15 movement, carried to a large extent by fast-food workers, has since seen successes on the state and local level. In California, for example, the minimum wage has been raised in stages since 2016 to reach the $15 threshold in 2022. At the same time, employment in the fast-food sector continues to go up and the industry faces a labor shortage. "Flippy," the burger-grilling robot featured in the *Wall Street Journal* commentary has indeed made it into several restaurants of the chain CaliBurger, where it has become a customer attraction. Flippy, a robot arm with a spatula attached, uses machine vision and thermal sensors to decide at which point it needs to flip the patty. Other than these capabilities, however, it needs a lot of help from its human coworkers: They need to place the patties on the grill, add cheese and other toppings before wrapping it and handing it over to the customers.[19]

TOWARD THE END OF LABOR

Today's world is still a world of labor. An overwhelming majority of the globe's population continues to spend a great deal of their waking time working. In all its dimensions, work is still an extraordinary factor in constituting and stratifying societies. All progress in and all speculation around digitally driven automation notwithstanding, it seems unlikely that this fact will change in the near future. Both the traditional factory and the digital factory of today stand in a complex and equivocal relation to processes and discourses of automation. "From the beginning of industrialization, the history of the factory was linked with the vision of automation, which ultimately aimed at the idea of a factory without people," writes Karsten Uhl, a historian of the factory, only to add that it is "characteristic of this way of thinking, however, that from the first automated spinning machines, to Taylorism, to the numerically controlled . . . machines of the post-war period, the possibilities for automation brought by innovations were always overestimated."[20]

Clearly, digital technology is at the core of a new circuit of automation that already has profound effects and will change and eliminate even more jobs in the future. Just as clearly, however, one can compare the current hype about robots and AI eliminating jobs with historical examples of automation discourses starting at least in the nineteenth century with authors such as Charles Babbage and Andrew Ure, whose visions of fully automated factories were influential for Marx and other contemporaries. Those visions of the imminent end of human labor would return periodically in the 1930s, 1950s, 1980s, and again in recent years, as Aaron Benanav points out in his *Automation and the Future of Work*.[21] All of these upticks in the discourse on automation were bound up with real processes of automation that replaced jobs and made workers redundant. After each wave of automation, however, more people than ever before had entered into wage labor.

The fact that these earlier waves of automation did not erase millions of jobs does not mean it cannot proceed differently this time, especially as the current situation is already characterized by a global oversupply of labor. Still, there is reason to remain skeptical about the end of labor. As Benanav points out convincingly, the current phenomenon of global underemployment might be due more to slowing growth and productivity rates than to leaps in productivity caused by digital technology.[22] At the same time, the broad rollout of automation technology is not simply a question of technological development, but also economic calculation, and thus it always exists in price competition with human labor. Also, digital technology can automate many tasks but generates at the same time new tasks and problems that require human labor. Statistics that give automation forecasts by sector and country and to the percentage point are good for media headlines, but they are rough estimations at best. This book has mostly steered clear of such debates on possible futures of automation and engaged with already existing forms of automation, or rather the changing relation of technology and living labor in the contemporary moment. This perspective does not allow statistical projections on the job market impact of digital automation. It allows, however, for empirical and theoretical points of skepticism against such linear and clear-cut projections and models.

Across the different sites of this book's investigations, both the impact and power of digital technology as well as the persistence of human labor became visible. Amazon's distribution centers are again a case in point: in recent years, Amazon has introduced hundreds of thousands of robots moving shelves automatically through the distribution centers. In the same time frame, it has also hired an even greater number of new workers who work in unison with these robots. The tiresome labor of crowdworkers training machine learning applications and content moderators deleting unwanted content from social media gives us more examples, both of the current limits of and the labor behind automation. These two professions are also exemplary for workforces hidden behind the supposed magic of algorithms. There are numerous examples in very different sectors for human workers behind ostensibly automated processes or AI applications that train software, evaluate its work, help in difficult situations, or are flat-out masked as algorithmic applications when there is only human labor. While there are stunning examples of the capability of machine learning algorithms and sophisticated robots, there are also funny examples of algorithms misunderstanding the simplest commands and sobering setbacks in robotics that show that digitally driven automation will fall short of replacing human labor even in some of the sectors thought to be easily automatable for quite some time. At this writing, the rise of algorithmic management, new forms of control and measurement of labor, new labor geographies enabled by digital technology, new gendered and racialized divisions of labor, new social polarizations, and the rise of contingent and flexible labor are far more impactful than the loss of jobs to robots.

The question of automation is still an important issue, especially because it forces us to think into the future. In many cases, not only unions but many forms of labor movements have committed to defending labor against automation and new technology. Concerning the often-catastrophic consequences of job losses, this makes a lot of sense. Observed from a distance, however, it is among the greatest absurdities of this social formation: the fact that technology can replace labor is seen as a threat, particularly by those who labor in the hardest jobs, with the longest hours, in the most precarious conditions and for the lowest wages.

This invokes the social questions bound up with automation and shows that the question of advancing alternative visions of the future is a pressing one. It is a question of destabilizing Fordist nostalgia expressed in the longing for the return of a class compromise, which was never particularly inclusive to begin with, and whose social and economic conditions have since evaporated. The constellations that allow visions of automation for the common good to become powerful are very much questions of political power, but they also point to how to approach the question of technology. Throughout this book, the various algorithmic infrastructures appeared predominantly as a means of organizing and controlling labor, as technologies to accelerate circulation and increase productivity. Technology is materialized social labor and is as such a product of the social relations that brought it into existence. It is thus little surprise that much contemporary technology is designed first and foremost to appropriate and privatize the labor of others, both that of individual workers and that of broader social cooperation. However, this does in no way mean that it has to stay that way.

7

THE CONTAGIOUS FACTORY

Epilogue

Writing a book about the digital transformation of global capitalism has sometimes felt like an impossible task. The mere speed and dynamic with which this process is occurring seem to be at odds with the long duration of publishing a book. All of the sites investigated in this book are part of dynamic transformations and were sometimes changing under my eyes during the time the research took place. In writing such a book, you always run the risk that some things may look different at the time the book is published; sometimes, you end up envying the historian researching ancient events. In 2020, these risks and feelings were radically amplified. The bulk of the manuscript was finished a few weeks before the coronavirus went global and took hold of my hometown, Berlin. At the time I am writing these words, late in 2020, a second wave of the virus is surging in many regions of the world, and the German government has mandated a second lockdown. At this moment, it is hard to tell when and if this pandemic will lose its momentum. At the same time, it seems clear that the conditions that produced this virus and its global implications will cause more events like the current pandemic in the future. Described by many as the Anthropocene, or maybe more aptly, the Capitalocene, the last century has dramatically changed the globe and the climate (and the factory has played no small role in this development).[1] This multifaceted ecological destruction played a role in the current pandemic and seems destined to cause future catastrophes. All uncertainty about the future aside, it is clear that the COVID-19 pandemic has already profoundly changed our societies and will change them even more in the future, although precisely how these changes will play out is rather unclear at this moment.

Reflecting on the book in light of the pandemic, it still does not feel like events have overtaken this book and its conclusions. Quite the contrary, many of the processes described in the chapters have been accelerated by the pandemic. Amazon, for example, is among the biggest profiteers of the crisis. The company has hired hundreds of thousands of workers in its warehouses and on the last mile to satisfy globally increased demand for the delivery of goods to the safety of consumers' homes. Since the onset of the crisis the company's stock price—and the wealth of Jeff Bezos—has risen. In many ways, platforms came to the spotlight in the current crisis. In China, the COVID-19 crisis led to a boom for platforms that deliver meals and other shopping to the doors of their isolated clients. In Paris and Milan, riders for the food-delivery platforms were often the only people to still be seen on the empty street throughout the dramatic first lockdowns of the cities. The pandemic shows in many ways the extent to which platforms have evolved into infrastructures of everyday life, a development that often coincides with a lack of public infrastructure—another thing that has become evident in the wake of COVID-19.

The riders of the delivery platforms, the workers in Amazon's distribution centers, and many more so-called frontline workers fulfill important social functions amid the crisis and are thereby often exposed to particular risks. Tens of thousands of warehouse workers contracted the virus in the first months of the pandemic, and the gig workers around the world—often migrants—faced the precarity of jobs with very little security such as continued payment of wages in the event of business closures, curfews, or sickness. Many therefore continue to work despite fears of the virus and dwindling business. "I can't self-quarantine because not working is not an option," writes Mariah Mitchell, who works for a variety of platforms such as Lyft and Uber Eats in New York and also is involved in a campaign for better working conditions in the gig economy. In a vivid letter that the *New York Times* published early in the pandemic, she explained: "If I don't make enough money, I can't feed my children for the next six weeks. I'm not stopping, fever or no fever. And that's what most other gig workers would do too, because none of us makes enough money to save up for an emergency like this."[2] Self-isolation and the home office are luxuries many cannot afford.

Those people who were indeed sent to work in the home office by their employers came upon an army of other platform workers for whom working from home has been a reality for a long time: the workers on online labor platforms who have been working from their kitchens and bedrooms for many years. Such crowdwork platforms might be another winner of the COVID-19 crisis, and the logic of remote work on demand could proliferate even more. It is equally easy to see how online games might be a refuge for both people seeking distraction from a world shaken up by the virus and masses of newly unemployed searching for ways to make any income.[3]

When Facebook sent the majority of its workers home in March 2020, many of the platforms' users complained that their harmless postings were marked as spam or dangerous and deleted from the platform while malicious content stayed up. The reason for these mistakes was simple: with the majority of the tens of thousands of workers that do the content moderation for Facebook out of their offices (and mostly unable to work out of their homes due to privacy concerns), the platform tried to substitute these workers with automated systems. The experience of many users, as well as the fact that Facebook's contractor Accenture required its content moderators to return to the office in the middle of exploding virus numbers in October 2020, shows once more that there are many problems artificial intelligence still cannot solve. The mandatory move back to the office was met by Accenture's subcontracted workers with worry and demands for better pay for their risk. Content moderators for other platforms also returned to their offices, while the regular employees of Facebook and other companies enjoy greatly enlarged and often permanent options of remote work.[4]

Just as with content moderation, the crisis showcases both the continuing importance of human labor and increased efforts of automation to ensure the uninterrupted flow of production in times of corona. The crisis might indeed be the starting point for renewed efforts at automation and lay the groundwork for more changes in the global economy. While these efforts will surely be of mixed success in the future, the crisis has already had dramatic effects on labor in the present. Since the onset of the crisis, millions of people have lost their jobs and will join the ranks of the un- and underemployed. Just as the last big crisis, the 2008 financial crisis, gave birth to the gig economy as we know it, this crisis may further spread and normalize contingent forms of labor. Growing numbers of unemployed workers and informal work arrangements also exert pressure on those workers holding onto more or less regular forms of employment and weaken their position vis-à-vis their employers. These developments will probably be more important issues for labor in the coming years than job loss through automation. It is also the effects of these developments, especially in times of collapsing health systems and public infrastructures with further austerity measures on the horizon, that provide the framework for the coming social conflicts over who will pay the price of this crisis and how our societies will be reorganized.

ACKNOWLEDGMENTS

This project took many years, and I am indebted to a great number of people without whom it would not have been possible. Many individuals and collectives have contributed greatly through their encouragement, critique, cooperation, discussion, and other forms of support and friendship. I want to express my deep gratitude for many very different things to Fabian Altenried, Franz Hefele, Julia Dück, Louisa Prause, Manuela Bojadžijev, Mariana Schütt, Mira Wallis, Morten Paul, Sabrina Apicella, Samira Spatzek, Sibille Merz, Svenja Bromberg, Theresa Hartmann, Ulrike Altenried, Verena Namberger, and many more.

The origins of this research lie in my work at Goldsmiths, University of London. I would like to thank Scott Lash and Alberto Toscano for all their support; Tiziana Terranova and Matteo Mandarini; and Yari Lanci, Luciana Parisi, and many others who made my time at Goldsmiths something I like to remember. My work would not have been possible without the financial support I received from the Rosa Luxemburg Foundation and the ESRC Doctoral Training Centre at Goldsmiths and Queen Mary.

Many collective forms of research and discussion in London, Berlin, Lüneburg, and other places shaped this project. I would like to especially thank Sandro Mezzadra, Brett Neilson, and Ned Rossiter for the various forms of collaboration that have enriched this project. Among the important collaborations with them and many others were the summer schools Investigating Logistics and Expanding the Margins and the network of scholars and activists involved in these projects, and the Politics and Infrastructures of Mobile Labor research project at the Humboldt University of Berlin. Other projects aiming at the collective production of knowledge included the reading group and conference Crisis and Critique of the State at Goldsmiths. Important newer projects and collective endeavors include especially the project Digitalisation of Labour and Migration with Manuela Bojadžijev and Mira Wallis and other projects with the rest of the Cabin Crew.

A special thank goes to Elizabeth Branch Dyson at the University of Chicago Press for her early enthusiasm for the project as well as her gracious support and guidance throughout the whole process. I also thank Mollie McFee for her support in the process of publishing the book and the two anonymous reviewers for their helpful comments.

Some sections and variations of arguments of the book were published with other outlets. A piece on the role of logistics in global capitalism was published in the French journal *Période*, and some considerations on last-mile delivery in the journal *Work Organisation, Labour & Globalisation*. Parts of chapter 3 flowed into an article cowritten with Manuela Bojadžijev, and some of the arguments and short variations in chapter 4 appeared in the journals *Prokla* and *Capital & Class* (see Altenried, "Le container et l'algorithme," "Die Plattform als Fabrik," "On the Last Mile," and "The Platform as Factory"; and Altenried and Bojadžijev, "Virtual Migration, Racism and the Multiplication of Labour").

Finally, I would like to thank the workers, trade unionists, activists, and other interviewees for their time—I can only hope that I was able to give something back.

NOTES

CHAPTER ONE

1. Andrew Norman Wilson, *Workers Leaving the Googleplex*, 2011, http://vimeo.com/15852288.
2. Google, *Google Interns' First Week*, 2013, https://www.youtube.com/watch?v=9No-FiEInLA.
3. Entry by user identifying as former scanning worker on an online job review website, August 2013.
4. Freeman, *Behemoth*, xvii.
5. What is a factory? It is generally understood to be a large building designed for production involving a greater number of workers. In the attempt to the define particularity of the factory, however, large-scale and (partly) automated machinery is central (see Gorißen, "Fabrik"; Uhl, "Work Spaces"). One driving force for the development of modern factories was machinery too large to be housed in private homes or small workshops. In many approaches, however, the crucial point is not the size of the machinery but rather how its starts to structure and dominate the production process (see Tronti, *Arbeiter und Kapital*, 28; Marx, *Capital*, vol. 1, 544). This includes an inversion of the role of living labor and technology, "in handicrafts and manufacture, the worker makes use of a tool; in the factory, the machine makes use of him," as Marx formulates it succinctly (Marx, *Capital*, vol. 1, 548). In his thinking, the factory is the central architectural form of capitalism and the paradigmatic space of real subsumption. While in Marx, and most of the eighteenth and first part of the nineteenth century, the factory as a building is considered not much more than a shell for the production process, this changes toward the end of the nineteenth century. In the efforts to build a "rational factory," the factory becomes more than a building; it becomes an essential component of the production process, or "the master machine" (see Biggs, *The Rational Factory*).
6. Some examples: Looking at education and the changing labor market, Phillip Brown, Hugh Lauder, and David Ashton use the term *digital Taylorism*

to describe the "industrialization of knowledge work" (see Brown, Lauder, and Ashton, *The Global Auction*, 74). David Noble also looks at education in his *Digital Diploma Mills*. Labor scholar Simon Head also stresses the effect of software to intensify work and deskill workers, arguing that we live in a "new age of 'scientific management'" (see Head, *The New Ruthless Economy*, 6). A research project on IT manufacturing diagnoses a "flexible Neo-Taylorism" (see Hürtgen et al., *Von Silicon Valley nach Shenzhen*, 274), a study on service labor finds a "subjectivized Taylorism" (see Matuschek, Arnold, and Voß, *Subjektivierte Taylorisierung*), while still other studies on warehouse labor speak of a new Taylorism (see Butollo et al., "Wie Stabil Ist der Digitale Taylorismus?"; Lund and Wright, "State Regulation and the New Taylorism"; Nachtwey and Staab, "Die Avantgarde des Digitalen Kapitalismus").

7. *The Economist*, "Digital Taylorism."
8. Mezzadra and Neilson, *Border as Method* and *The Politics of Operations*.
9. Easterling, *Extrastatecraft*.
10. Toscano, "Factory, Territory, Metropolis, Empire," 200.
11. On these very material infrastructures of the digital, see, for example, Gabrys, *Digital Rubbish*; Hu, *A Prehistory of the Cloud*; Mosco, *To the Cloud*; Parks and Schwoch, *Down to Earth*; Parks and Starosielski, *Signal Traffic*; Starosielski, *The Undersea Network*.
12. For some important perspectives in software studies, see, for example, Dodge and Kitchin, *Code/Space*; Fuller, *Software Studies*; Fuller and Goffey, *Evil Media*; Parisi, *Contagious Architecture*; Terranova, "Red Stack Attack!"
13. For an instructive overview on methods and problems in researching algorithms, mainly from the perspective of the social sciences, see Kitchin, "Thinking Critically about and Researching Algorithms."
14. Tsing, "Supply Chains and the Human Condition," 150, 148.
15. Harvey, *Spaces of Capital*, 121.
16. See, for example, Lazzarato, "Immaterial Labour"; Hardt and Negri, *Empire*.
17. See, for example, Virno, *A Grammar of the Multitude* and "General Intellect."
18. Negri, *Goodbye Mr. Socialism*, 114, translation amended.
19. See, for example, Caffentzis, *In Letters of Blood and Fire*; Dyer-Witheford, *Cyber-Marx*, "Empire, Immaterial Labor," and *Cyber-Proletariat*; Huws, *The Making of a Cybertariat*, "Logged Labour," and *Labor in the Global Digital Economy*; Irani, "Justice for 'Data Janitors.'"
20. Braverman, *Labor and Monopoly Capital*; Tronti, *Arbeiter und Kapital*.

CHAPTER TWO

1. Marx, *Capital*, vol. 1, 125.
2. Reifer, "Unlocking the Black Box of Globalization."
3. Toscano and Kinkle, *Cartographies of the Absolute*, 201.
4. Thrift, *Knowing Capitalism*, 213.
5. Cowen, *The Deadly Life of Logistics*, 23.

6. Harney and Moten, *The Undercommons*, 110.
7. Bonacich and Wilson, *Getting the Goods*, 3.
8. Cowen, *The Deadly Life of Logistics*, 30, emphasis in original.
9. Cowen, 2.
10. When seeking to derive a theory of logistics (rather than simple transportation) from Marx's scattered remarks on the topic, particularly in the *Grundrisse* and volume 2 of *Capital*, logistics appears as both productive enterprise in its own right and as something difficult to conceptually delineate from the process of production. He answers his own question ("Can the capitalist valorise the road [*den Weg verwerten*]?" Marx, *Grundrisse*, 526, translation amended) in the affirmative. Changing the location of goods can be a commodity in its own right: "The 'circulating' of commodities, i.e., their actual course in space, can be resolved into the transport of commodities. The transport industry forms on the one hand an independent branch of production, and hence a particular sphere for the investment of productive capital" (Marx, *Capital*, vol. 2, 229). On the other hand, and perhaps more importantly, capital's expansionist logic engenders a process that makes it increasingly difficult to distinguish between production and physical circulation. To Marx, the "precondition of production based on capital is . . . *the production of a constantly widening sphere of circulation*" (Marx, *Grundrisse*, 407, emphasis in original). Accordingly, the "tendency to create the *world market* is directly given in the concept of capital itself. Every limit appears as a barrier to be overcome" (Marx, *Grundrisse*, 408, emphasis in original), as Marx's oft-quoted phrases claim. In a passage quoted less often, he goes on to argue that this process has the tendency "to subjugate every moment of production itself to exchange," until finally "commerce no longer appears here as a function taking place between independent productions for the exchange of their excess but, rather, as an essentially *all-embracing presupposition and moment of production itself*," Marx, *Grundrisse*, 408, emphasis added; see also Altenried, "Le container et l'algorithme."
11. Cowen, *The Deadly Life of Logistics*, 2, emphasis in original.
12. Rossiter, *Software, Infrastructure, Labor*.
13. According to a company publication titled "SAP: The World's Largest Provider of Enterprise Application Software. SAP Corporate Fact Sheet, 2017," https://www.sap.com/corporate/en/documents/2016/07/0a4e1b8c-7e7c-0010-82c7-eda71af511fa.html.
14. Campbell-Kelly, *From Airline Reservations to Sonic the Hedgehog*, 197.
15. See Rossiter, *Software, Infrastructure, Labor*.
16. Cowen, *The Deadly Life of Logistics*, 23.
17. Holmes, "Do Containers Dream of Electric People?," 41.
18. On the history and present of Walmart, see also Brunn, *Wal-Mart World*; LeCavalier, *The Rule of Logistics*; Lichtenstein, *The Retail Revolution*.
19. Recode, "Jeff Bezos vs. Peter Thiel and Donald Trump | Jeff Bezos, CEO Amazon | Code Conference 2016," 2016, https://www.youtube.com/watch?v=guVxubbQQKE&.

20. "Amazon zieht positive Zwischenbilanz der Weihnachtssaison," press release, December 20, 2013, https://amazon-presse.de/Logistikzentren/Logistikzentren-in-Deutschland/Presskit/amazon/de/News/Logistikzentren/download/de/News/Logistikzentren/Amazon-zieht-positive-Zwischenbilanz-der-Weihnachtssaison.pdf/.
21. Interview with Amazon Brieselang worker on the fringes of a political meeting, October 2015.
22. Interview with dismissed Amazon Brieselang worker, Berlin, March 2015.
23. Interview with Amazon Brieselang worker, Berlin, May 2015.
24. On the history of the barcode, see, for example, Nelson, *Punched Cards to Bar Codes*.
25. Quoted in Dodge and Kitchin, "Codes of Life," 859.
26. "Time-based warehouse movement maps," US Patent 7243001 B2 2007, Amazon Technologies, Inc., Janert et al., https://www.google.com/patents/US7243001.
27. Lyster, *Learning from Logistics*, 3.
28. Conversation with Amazon Leipzig worker on the fringes of a political networking event Berlin, December 2015.
29. Conversation with another Amazon Leipzig worker on the fringes of a networking event in Berlin, December 2015.
30. Interview with Amazon Brieselang worker, Berlin, May 2015.
31. Interview with Amazon Brieselang worker, Berlin, December 2014.
32. LeCavalier, *The Rule of Logistics*, 152.
33. Amazon manager, as quoted in the *Financial Times*, see O'Connor 2013.
34. Marx, *Grundrisse*, 692.
35. "Inactivity Protocol" issued in 2014 and made public by the union ver.di, https://www.amazon-verdi.de/4557. My translation.
36. Diana Löbl and Peter Onneken, ARD, "Ausgeliefert! Leiharbeiter bei Amazon," aired February 13, 2013.
37. According to the union Inicjatywa Pracownicza; see, for example, the blog *LabourNet*: "Amazon im Weihnachtsstress—Das Warenlager in Poznan," April 15, 2015, http://www.labournet.de/internationales/polen/arbeitsbedingungen-polen/amazon-im-weihnachtsstress-das-warenlager-in-poznan/.
38. Report on the 2016 Christmas season at Amazon FC in Hemel Hempstead, posted anonymously on the website Angry Workers of the World, January 7, 2017, https://angryworkersworld.wordpress.com/2017/01/17/calling-all-junglists-a-short-report-from-amazon-in-hemel-hempstead/.
39. Amazon, "CamperForce. Your next adventure is here," online job listing on Amazon (n.d.), http://www.amazondelivers.jobs/about/camperforce/, accessed April 22, 2017.
40. Boewe and Schulten, *The Long Struggle of the Amazon Employees*, 39.
41. See also the important work of Sabrina Apicella on strikes at Amazon (Apicella, *Amazon in Leipzig*).
42. "Airborne fulfillment center utilizing unmanned aerial vehicles for item

delivery," US Patent 9305280 granted 2016 to Amazon Technologies, Inc., Berg et al. https://www.google.com/patents/US9305280.

43. Edwin Lopez, "Why Is the Last Mile so Inefficient?" *Supply Chain Dive* (blog), May 22, 2017, https://www.supplychaindive.com/news/last-mile-spotlight-inefficient-perfect-delivery/443089/.
44. Lyster, *Learning from Logistics*, 13.
45. Lyster, 3.
46. UPS SEC filing for 2017, February 2018, http://www.investors.ups.com/static-files/8d1241ae-4786-42e2-b647-bf34e2954b3e.
47. Allen, "The UPS Strike, 20 Years Later."
48. Bruder, "These Workers Have a New Demand."
49. Jacob Goldstein, "To Increase Productivity, UPS Monitors Drivers' Every Move," NPR, April 17, 2014, https://www.npr.org/sections/money/2014/04/17/303770907/to-increase-productivity-ups-monitors-drivers-every-move.
50. "Telematics," UPS Leadership Matters website, https://www.ups.com/content/us/en/bussol/browse/leadership-telematics.html, accessed July 9, 2018.
51. Frank, "How Telematics Has Completely Revolutionized the Management of Fleet Vehicles."
52. Burnett, "Coming Full Circle."
53. "ORION: The Algorithm Proving That Left Isn't Right," *UPS Compass* (blog), October 5, 2016, https://www.ups.com/us/en/services/knowledge-center/article.page?kid=aa3710c2.
54. Post in an independent online forum run by UPS workers, January 2016.
55. Kaplan, "The Spy Who Fired Me."
56. Bonacich and Wilson, *Getting the Goods*, 113.
57. Smith, Marvy, and Zerolnick, *The Big Rig Overhaul*.
58. Entry by user identifying as an Amazon Flex driver on an online job review website, March 2017.
59. *Lawson v. Amazon Inc.*, case filed at United States District Court of California, 2017, http://www.courthousenews.com/wp-content/uploads/2017/04/Amazon.pdf.
60. Entry by user identifying as an Amazon Flex driver on an online job review website, April 2017.
61. Sekula, *Fish Story*, 12.
62. Cuppini, Frapporti, and Pirone, "Logistics Struggles in the Po Valley Region."

CHAPTER THREE

1. *Hernandez v. IGE*, case filed at the US District Court Southern Florida, 2007, https://dockets.justia.com/docket/florida/flsdce/1:2007cv21403/296927.
2. Heeks, "Current Analysis and Future Research Agenda on 'Gold Farming.'"

3. "I am payed to play MMORPGs and it sucks." Anonymous post on Cracked.com, April 16, 2016, http://www.cracked.com/personal-experiences-2228-im-paid-to-play-mmorpgs-its-nightmare-5-realities.html.
4. Gold farming worker cited in the *New York Times* (see Barboza, "Boring Game? Hire a Player").
5. Dibbell, The Chinese Game Room."
6. Gold farmer, as cited in the blog *Eurogamer*, entry by Nick Ryan, March 25, 2009, http://www.eurogamer.net/articles/gold-trading-exposed-the-sellers-article?page=3.
7. Gold farmer, as cited in the *South China Morning Press* (see Huifeng, "Chinese 'Farmers' Strike Cyber Gold").
8. Rettberg, "Corporate Ideology in World of Warcraft," 30.
9. Quote from MMOGA.de, https://www.mmoga.com/content/Intermediation-Process.html, accessed September 13, 2014.
10. According to the blog *Online Marketing Rockstars*, entry by Torben Lux, June 13, 2016, https://omr.com/de/exklusiv-mmoga-exit/.
11. Gold farm owner cited in the *New York Times* (see Barboza, "Boring Game? Hire a Player").
12. Employee at an intermediary platform, as cited in the blog *Eurogamer*, entry by Nick Ryan, March 25, 2009, http://www.eurogamer.net/articles/gold-trading-exposed-the-sellers-article?page=3.
13. See, for example, Dibbell, "The Decline and Fall of an Ultra Rich Online Gaming Empire"; Heeks, "Understanding 'Gold Farming' and Real-Money Trading."
14. Gold farm owner cited in the *South China Morning Press* (see Huifeng, "Chinese 'Farmers' Strike Cyber Gold").
15. Ge Jin, *Goldfarmers*, parts 1–3. September 20, 2010, https://www.youtube.com/watch?v=q3cmCKjPLR8, https://www.youtube.com/watch?v=3rezLLMhwSM&t=85s, https://www.youtube.com/watch?v=kCXZNA74iIo.
16. Lao Liu, worker at a gold farm featured in *Goldfarmers*.
17. Another worker at a gold farm featured in *Goldfarmers*.
18. North American gamer and antifarming activist featured in *Goldfarmers*.
19. North American gamer and antifarming activist featured in *Goldfarmers*.
20. Nakamura, "Don't Hate the Player, Hate the Game."
21. Nakamura.
22. Aneesh, *Virtual Migration*.
23. Aneesh, 2.
24. Discussion on Reddit, https://www.reddit.com/r/2007scape/comments/6xnfso/killing_venezuelans_at_east_drags_guide/, accessed October 30, 2020.
25. As cited on the website Kotaku, entry by Nathan Grayson, April 2, 2018, https://www.kotaku.com.au/2017/10/the-runescape-players-who-farm-gold-so-they-dont-starve-to-death/.
26. Name of the company has been changed.

27. Interview with QA worker at Smalline Berlin who founded the works council, Berlin, December 2013.
28. Interview with QA worker at Smalline Berlin who founded the works council, Berlin, December 2013.
29. Interview with QA worker at Smalline Berlin who founded the works council, Berlin, December 2013.
30. Name of the company has been changed.
31. Telephone interview with union secretary (ver.di) responsible for Supgame Studios, Berlin / Hamburg, May 2017.
32. Telephone interview with union secretary (ver.di) responsible for Supgame Studios, Berlin / Hamburg, May 2017.
33. Name changed.
34. Conversation with worker at Smalline Berlin who founded the works council on the fringes of an office visit, Berlin, April 2014.
35. Conversation with worker at Smalline Berlin on the fringes of an office visit, Berlin, April 2014.
36. Interview with QA worker at Smalline Berlin who founded the works council, Berlin, December 2013.
37. Interview with QA worker at Smalline Berlin who founded the works council, Berlin, December 2013.
38. Conversation with worker at Smalline Berlin who founded the works council on the fringes of an office visit, Berlin, April 2014.
39. Telephone interview with ver.di union secretary responsible for Supgame Studios, Berlin / Hamburg, May 2017.
40. Interview with QA worker at Smalline Berlin who founded the works council, Berlin, December 2013.
41. See also Bulut, "Playboring in the Tester Pit."
42. Former QA worker cited in *Jacobin* (see Williams, "You Can Sleep Here All Night").
43. Interview with worker at Smalline Berlin who founded the works council, Berlin, December 2013.
44. Woodcock, "The Work of Play."
45. Conversation with worker at Smalline Berlin on the fringes of an office visit, Berlin, April 2014.
46. See, for example, Crogan, *Gameplay Mode*; Dyer-Witheford and De Peuter. *Games of Empire*; Kline, Dyer-Witheford, and De Peuter, *Digital Play*.
47. Kline, Dyer-Witheford, and De Peuter, *Digital Play*, 87–88.
48. Dyer-Witheford and De Peuter. *Games of Empire*, 27.
49. Dyer-Witheford and De Peuter. *Games of Empire*, 55.
50. The author of the letter created the blog *EA Spouse*, where the letter can be found, entry on November 10, 2004, http://ea-spouse.livejournal.com/274.html.
51. *EA Spouse*, entry on November 10, 2004, http://ea-spouse.livejournal.com/274.html.
52. Guth and Wingfield, "Workers at EA Claim They Are Owed Overtime."

53. Berry, *The Philosophy of Software,* 39.
54. Parikka, "Cultural Techniques of Cognitive Capitalism," 42.
55. Thompson, Parker, and Cox, "Interrogating Creative Theory and Creative Work."
56. Cited in Thompson, Parker, and Cox, 324.
57. Interview with worker at Smalline Berlin, office visit, Berlin, April 2014.
58. Ruffino and Woodcock, "Game Workers and the Empire."
59. Entry on Nicholas Carr's blog, December 5, 2006, http://www.roughtype.com/?p=611.

CHAPTER FOUR

1. Nick Masterton, "Outsourcing Offshore," 2013, https://vimeo.com/101622811.
2. MIT TechTV, "Opening Keynote and Keynote Interview with Jeff Bezos," 2006, http://techtv.mit.edu/videos/16180-opening-keynote-and-keynote-interview-with-jeff-bezos.
3. Terranova, "Free Labor"; Scholz, *Digital Labor.*
4. Irani, "Difference and Dependence among Digital Workers," 226.
5. MIT TechTV, "Opening Keynote and Keynote Interview with Jeff Bezos."
6. Kuek et al., *The Global Opportunity in Online Outsourcing,* 7; Heeks, "Decent Work and the Digital Gig Economy" and ""How Many Platform Workers Are There in the Global South?"
7. MIT TechTV, "Opening Keynote and Keynote Interview with Jeff Bezos."
8. Mechanical Turk, https://www.mturk.com/, accessed December 30, 2019.
9. Random sample HITs available on mturk.com, accessed December 3, 2016.
10. Clickworker, https://www.clickworker.com/, accessed December 30, 2019.
11. Appen, https://appen.com/solutions/training-data/, accessed October 23, 2020.
12. See also Schmidt, *Crowdproduktion von Trainingsdaten.*
13. Gray and Suri, *Ghost Work.*
14. Aytes, "Return of the Crowds."
15. Interview with Daniel, student and crowdworker, Berlin, March 2016.
16. Interview with Daniel, student and crowdworker, Berlin, March 2016.
17. Interview with Daniel, student and crowdworker, Berlin, March 2016.
18. Fang, ""Google Hired Gig Economy Workers."
19. MTurk requester cited by Lilly Irani (see Irani, "Difference and Dependence among Digital Workers," 228–29).
20. Irani, "Difference and Dependence among Digital Workers," 230–231.
21. Marx, *Capital,* vol. 1, 698.
22. Marx, *Capital,* vol. 1, 695.
23. Berg et al., *Digital Labour Platforms and the Future of Work.*
24. Post on We Are Dynamo, forum for crowdworkers that launched a letter-writing campaign to Jeff Bezos.
25. Post on We Are Dynamo.

26. Media report on the company, published on its website until mid-2019.
27. Altenried and Wallis, “Zurück in die Zukunft.”
28. See also Berg et al., *Digital Labour Platforms and the Future of Work*.
29. Post on We Are Dynamo.
30. Post on We Are Dynamo.
31. Post on We Are Dynamo.
32. Marx, *Capital*, vol. 1, 591.
33. Interview with the CEO at CrowdGuru’s offices, Berlin, March 2016.
34. Grier, *When Computers Were Human*.
35. Light, “When Computers Were Women.”
36. Chun, *Programmed Visions*, 30.
37. Post on company-run forum for crowdworkers, March 2014.
38. Beerepoot, Kloosterman, and Lambregts, *The Local Impact of Globalization in South and Southeast Asia*.
39. Graham and Anwar, “The Global Gig Economy.”
40. Post on We Are Dynamo.
41. Easterling, *Extrastatecraft*, 123.
42. Kucklick, “SMS-Adler.”
43. Mark Zuckerberg, “Is Connectivity a Human Right?,” Facebook, 2014, https://www.facebook.com/isconnectivityahumanright.
44. Graham and Mann, “Imagining a Silicon Savannah?”
45. Graham, Hjorth, and Lehdonvirta, “Digital Labour and Development.”
46. Berg et al., *Digital Labour Platforms and the Future of Work*, 52.
47. Silberman and Irani, “Operating an Employer Reputation System.”

CHAPTER FIVE

1. Karim Amer and Jehane Noujaim, *The Great Hack*, 113 min., distributed by Netflix, 2019.
2. See, for example, Fuchs, *Digital Labor and Karl Marx*.
3. Jasmine Enberg, “Global Digital Ad Spending 2019,” Emarketer, March 28, 2019, https://www.emarketer.com/content/global-digital-ad-spending-2019.
4. Nieborg and Helmond, “The Political Economy of Facebook’s Platformization in the Mobile Ecosystem,” 199.
5. Easterling, *Extrastatecraft*, 13.
6. Gerlitz and Helmond, “The Like Economy.”
7. Job listing on the company’s website, https://careers.lionbridge.com/jobs/rater-united-states, accessed March 20, 2020.
8. Matt McGee, “An Interview with a Search Quality Rater,” Search Engine Land, January 20, 2012, http://searchengineland.com/interview-google-search-quality-rater-108702.
9. McGee.
10. Pasquinelli, “Google’s PageRank Algorithm,” 153.
11. The speech was later published on the *Facebook for Developers* blog: Bret

Taylor, "The Next Evolution of Facebook Platform," *Facebook for Developers*, April 21, 2010, https://developers.Facebook.com/blog/post/377.

12. Taylor, "The Next Evolution of Facebook Platform."
13. Nechushtai, "Could Digital Platforms Capture the Media through Infrastructure?"
14. Quoted from Facebook's SEC statement, February 1, 2012, p. 90, https://www.sec.gov/Archives/edgar/data/1326801/000119312512034517/d287954ds1.htm.
15. Mosco, *To the Cloud*, 32.
16. See, for example, the research project "Data Farms. Circuits, Labour, Territory" with many current and upcoming publications by its participants; https://www.datafarms.org, accessed October 20, 2020.
17. Rossiter, *Software, Infrastructure, Labor*, 138.
18. Glanz, "Power, Pollution and the Internet."
19. "Artificial Reef Datacenter," Patent Application, Microsoft Technology Licensing, December 29, 2016, http://patentyogi.com/wp-content/uploads/2017/01/US20160381835.pdf.
20. According to *Fortune* (see Elegant, "The Internet Cloud Has a Dirty Secret").
21. Jones, "How to Stop Data Centres from Gobbling up the World's Electricity."
22. Greenpeace, *Make IT Green*.
23. Neilson and Notley, "Data Centres as Logistical Facilities."
24. See Bratton, *The Stack*; Rossiter, *Software, Infrastructure, Labor*.
25. Cited in the *New York Times* (see Satariano, "How the Internet Travels across Oceans").
26. Starosielski, *The Undersea Network*.
27. Burgess, "Google and Facebook's New Submarine Cable."
28. Satariano, "How the Internet Travels across Oceans."
29. Ngai and Ruckus, *iSlaves*; Ngai, *Migrant Labor in China*.
30. According to a worker interviewed in *Business Insider* (see Jacobs, "Inside 'iPhone City'").
31. Ngai, *Migrant Labor in China*, 127.
32. Easterling, *Extrastatecraft*, 36.
33. According to a press release by the World Bank, "World Bank Report Provides New Data to Help Ensure Urban Growth Benefits the Poor," January 26, 2015, http://www.worldbank.org/en/news/press-release/2015/01/26/world-bank-report-provides-new-data-to-help-ensure-urban-growth-benefits-the-poor.
34. Lüthje, *Standort Silicon Valley*; Pellow and Park, *The Silicon Valley of Dreams*.
35. Dyer-Witheford, *Cyber-Proletariat*, 71.
36. Lüthje, *Standort Silicon Valley*.
37. Hürtgen et al., *Von Silicon Valley nach Shenzhen*, 274.
38. Andrijasevic and Sacchetto, "Disappearing Workers."
39. Interview with Roberto (name changed), Arvato worker, Berlin, March 2019.
40. Interview with Roberto (name changed), Arvato worker, Berlin, March 2019.
41. Interview with Roberto (name changed), Arvato worker, Berlin, March 2019.

42. The leaked manual for content moderators working for Facebook through the crowdworking platform oDesk can be found online: https://de.scribd.com/doc/81877124/Abuse-Standards-6-2-Operation-Manual, accessed October 30, 2020.
43. According to an investigation by the German NGO Netzpolitik (see Dachwitz and Reuter, "Warum Künstliche Intelligenz Facebooks Moderationsprobleme").
44. Self-presentation on the company's website; https://www.arvato.com/en/about/facts-and-figures.html, accessed February 9, 2017.
45. Interview with Roberto (name changed), Arvato worker, Berlin, March 2019.
46. Cited in *The Verge* (see Newton, "The Terror Queue").
47. Newton, "The Terror Queue."
48. Interview with Roberto (name changed), Arvato worker, Berlin, March 2019.
49. Content moderator at Arvato Berlin, cited in the *Süddeutsche Zeitung* (see Grassegger and Krause, "Im Netz des Bösen").
50. See Grassegger and Krause, "Im Netz des Bösen."
51. Interview with Roberto (name changed), Arvato worker, Berlin, March 2019.
52. Roberts, *Behind the Screen*.
53. Report by Burcu Gültekin Punsmann, a former Arvato employee on her time as content moderator in Berlin published in the *Süddeutsche Zeitung* (see Punsmann, "Three Months in Hell").
54. Punsmann.
55. Punsmann.
56. Punsmann, translation amended.
57. Cited in *Vice* (see Koebler and Cox, "The Impossible Job").
58. According to an investigation by the German NGO Netzpolitik (see Reuter et al., "Exklusiver Einblick").
59. Cited in *Vice* (see Gilbert, ""Facebook Is Forcing Its Moderators to Log Every Second of Their Days").
60. Cited in Ciaran Cassidy and Adrian Chen, *The Moderators*, 2017, https://www.youtube.com/watch?v=k9m0axUDpro.
61. Cited in Cassidy and Chen.
62. Cited in Cassidy and Chen.
63. According to data collected by the World Bank; http://data.worldbank.org/indicator/NV.SRV.TETC.ZS?locations=PH, accessed January 23, 2017.
64. Bajaj, "A New Capital of Call Centers."
65. Quote from the website Microsourcing, https://www.microsourcing.com/why-offshore/why-the-philippines/, accessed January 19, 2020.
66. See also Roberts, "Digital Refuse."
67. Cited in the *Washington Post* (see Dwoskin, Whalen, and Cabato, "Content Moderators at YouTube, Facebook and Twitter").
68. Chen, "The Laborers Who Keep the Dick Pics and Beheadings out of Your Facebook Feed."
69. Helmond, Nieborg, and van der Vlist, "Facebook's Evolution: Development of a Platform-as-Infrastructure."

CHAPTER SIX

1. Rudacille, "In Baltimore, Visions of Life after Steel."
2. Loomis, "The Sinking of Bethlehem Steel."
3. Taylor, *The Principles of Scientific Management*, 41.
4. Freeman, *Behemoth*, 107.
5. According to a report by the *New York Times* on Amazon in Baltimore (see Shane, "Prime Mover").
6. Documents obtained by *The Verge*; see Lecher, "How Amazon Automatically Tracks and Fires Warehouse Workers."
7. Pias, "Computer Spiel Welten."
8. See also Raffetseder, Schaupp, and Staab, "Kybernetik und Kontrolle. Algorithmische Arbeitssteuerung und Betriebliche Herrschaft."
9. Barns, *Platform Urbanism*; Sadowski, "Cyberspace and Cityscapes."
10. Staab, *Digitaler Kapitalismus*.
11. Woodcock and Graham, *The Gig Economy*.
12. Bojadžijev and Karakayali, "Autonomie der Migration"; Mezzadra, "The Gaze of Autonomy."
13. Mezzadra and Neilson, *The Politics of Operations*; Altenried et al., "Logistical Borderscapes."
14. Mezzadra and Neilson, *The Politics of Operations*, 159.
15. Bojadžijev, "Migration und Digitalisierung."
16. Bruder, "Meet the Immigrants Who Took On Amazon."
17. Scholz and Schneider, *Ours to Hack and to Own*; Silberman and Irani, "Operating an Employer Reputation System."
18. *Wall Street Journal* opinion piece (see Puzder, "The Minimum Wage Should Be Called the Robot Employment Act").
19. According to Scott Neuman, "'Flippy' the Fast Food Robot (Sort Of) Mans the Grill at CaliBurger," NPR, March 5, 2018, https://www.npr.org/sections/thetwo-way/2018/03/05/590884388/flippy-the-fast-food-robot-sort-of-mans-the-grill-at-caliburger.
20. Uhl, "Work Spaces."
21. Benanav, *Automation and the Future of Work*.
22. Benanav.

CHAPTER SEVEN

1. Moore, *Anthropocene or Capitalocene?* and *Capitalism in the Web of Life*; Wallace, *Dead Epidemiologists*.
2. Mitchell, "'I Deliver Your Food"; see also Altenried, Bojadžijev, and Wallis, "Platform Im/mobilities"; Altenried, Niebler, and Wallis, "On-Demand. Prekär. Systemrelevant."
3. Dyer-Witheford and De Peuter, "Postscript."
4. See Statt, Newton, and Schiffer, "Facebook Moderators at Accenture Are Being Forced Back to the Office."

BIBLIOGRAPHY

Allen, Joe. "The UPS Strike, 20 Years Later." *Jacobin*, August 8, 2017. https://www.jacobinmag.com/2017/08/ups-strike-teamsters-logistics-labor-unions-work.

Altenried, Moritz. "Le container et l'algorithme: La Logistique dans le capitalisme global." *Revue Période*, February 11, 2016. http://revueperiode.net/le-container-et-lalgorithme-la-logistique-dans-le-capitalisme-global/.

Altenried, Moritz. "On the Last Mile: Logistical Urbanism and the Transformation of Labour." *Work Organisation, Labour & Globalisation* 13, no. 1 (2019): 114–29.

Altenried, Moritz. "The Platform as Factory: Crowdwork and the Hidden Labour behind Artificial Intelligence." *Capital & Class* 44, no. 2 (2020): 145–58.

Altenried, Moritz. "Die Plattform als Fabrik. Crowdwork, Digitaler Taylorismus und die Vervielfältigung der Arbeit." *PROKLA. Zeitschrift für kritische Sozialwissenschaft* 46, no. 2 (2017): 175–92.

Altenried, Moritz, and Manuela Bojadžijev. "Virtual Migration, Racism and the Multiplication of Labour." *Spheres: Journal for Digital Cultures*, June 23, 2017. http://spheres-journal.org/virtual-migration-racism-and-the-multiplication-of-labour/.

Altenried, Moritz, Manuela Bojadžijev, Leif Höfler, Sandro Mezzadra, and Mira Wallis. "Logistical Borderscapes: Politics and Mediation of Mobile Labor in Germany after the 'Summer of Migration.'" *South Atlantic Quarterly* 117, no. 2 (2018): 291–312.

Altenried, Moritz, Manuela Bojadžijev, and Mira Wallis. "Platform Im/mobilities: Migration and the Gig Economy in Times of Covid-19." *Routed: Migration & (Im)Mobility Magazine*, October 2020. https://www.routedmagazine.com/platform-immobilities.

Altenried, Moritz, Valentin Niebler, and Mira Wallis. "On-Demand. Prekär. Systemrelevant." *Der Freitag*, March 25, 2020. https://www.freitag.de/autoren/der-freitag/on-demand-prekaer-systemrelevant.

Altenried, Moritz, and Mira Wallis. "Zurück in die Zukunft: Digitale Heimarbeit." *Ökologisches Wirtschaften*, April 2018: 24–27.

Andrijasevic, Rutvica, and Devi Sacchetto. "'Disappearing Workers': Foxconn in Europe and the Changing Role of Temporary Work Agencies." *Work, Employment and Society* 31, no. 1 (2017): 54–70.

Aneesh, A. *Virtual Migration: The Programming of Globalization*. Durham, NC: Duke University Press, 2006.

Apicella, Sabrina. *Amazon in Leipzig. Von den Gründen (nicht) zu streiken*. Berlin: Rosa Luxemburg Stiftung, 2016.

Aytes, Ayhan. "Return of the Crowds: Mechanical Turk and Neoliberal States of Exception." In *Digital Labor: The Internet as Playground and Factory*, edited by Trebor Scholz, 79–97. London: Routledge, 2013.

Bajaj, Vikas. "A New Capital of Call Centers." *New York Times*, November 25, 2011. http://www.nytimes.com/2011/11/26/business/philippines-overtakes-india-as-hub-of-call-centers.html.

Barboza, David. "Boring Game? Hire a Player." *New York Times*, December 9, 2005. http://www.nytimes.com/2005/12/09/technology/boring-game-hire-a-player.html.

Barns, Sarah. *Platform Urbanism: Negotiating Platform Ecosystems in Connected Cities*. Singapore: Palgrave Macmillan, 2020.

Beerepoot, Niels, Robert Kloosterman, and Bart Lambregts. *The Local Impact of Globalization in South and Southeast Asia*. London: Routledge, 2015.

Benanav, Aaron. *Automation and the Future of Work*. London: Verso, 2020.

Berg, Janine, Marianne Furrer, Ellie Harmon, Uma Rani, and Six Silberman. *Digital Labour Platforms and the Future of Work. Towards Decent Work in the Online World*. Geneva: International Labour Office, 2018.

Berry, David. *The Philosophy of Software: Code and Mediation in the Digital Age*. Basingstoke, UK: Palgrave Macmillan, 2011.

Biggs, Lindy. *The Rational Factory: Architecture, Technology and Work in America's Age of Mass Production*. Baltimore: Johns Hopkins University Press, 1996.

Boewe, Jörn, and Johannes Schulten. *The Long Struggle of the Amazon Employees*. Brussels: Rosa Luxemburg Foundation, 2017.

Bojadžijev, Manuela. "Migration und Digitalisierung. Umrisse eines emergenten Forschungsfeldes." In *Jahrbuch Migration und Gesellschaft 2019/2020*, edited by Hans Karl Peterlini and Jasmin Donlic, 15–28. Bielefeld, Germany: Transcript, 2020.

Bojadžijev, Manuela, and Serhat Karakayali. "Autonomie der Migration: Zehn Thesen zu einer Methode." In *Turbulente Ränder: Neue Perspektiven auf Migration an den Grenzen Europas*, edited by Transit Migration, 203–9. Bielefeld, Germany: Transcript, 2007.

Bonacich, Edna, and Jake B. Wilson. *Getting the Goods: Ports, Labor, and the Logistics Revolution*. Ithaca, NY: Cornell University Press, 2008.

Bratton, Benjamin H. *The Stack: On Software and Sovereignty*. Cambridge, MA: MIT Press, 2016.

Braverman, Haryy. *Labor and Monopoly Capital: The Degradation of Work in the Twentieth Century*. New York: Monthly Review Press, 1998.

Brown, Phillip, Hugh Lauder, and David Ashton. *The Global Auction: The Broken Promises of Education, Jobs, and Incomes*. Oxford: Oxford University Press, 2012.

Bruder, Jessica. "Meet the Immigrants Who Took On Amazon." *Wired*, November 12, 2019. https://www.wired.com/story/meet-the-immigrants-who-took-on-amazon/.

Bruder, Jessica. "These Workers Have a New Demand: Stop Watching Us." *The Nation*, May 27, 2015. https://www.thenation.com/article/these-workers-have-new-demand-stop-watching-us/.

Brunn, Stanley D. *Wal-Mart World: The World's Biggest Corporation in the Global Economy*. New York: Routledge, 2006.

Bulut, Ergin. "Playboring in the Tester Pit: The Convergence of Precarity and the Degradation of Fun in Video Game Testing." *Television & New Media* 16, no. 3 (2015): 240–58.

Burgess, Matt. "Google and Facebook's New Submarine Cable Will Connect LA to Hong Kong." *Wired*, April 6, 2017. http://www.wired.co.uk/article/google-facebook-plcn-internet-cable.

Burnett, Graham D. "Coming Full Circle." *Cabinet*, no. 47 (2012): 73–77.

Butollo, Florian, Thomas Engel, Manfred Füchtenkötter, Robert Koepp, and Mario Ottaiano. "Wie Stabil Ist der Digitale Taylorismus? Störungsbehebung, Prozessverbesserungen und Beschäftigungssystem bei einem Unternehmen des Online-Versandhandels." *AIS-Studien* 11, no. 2 (2018): 143–59.

Caffentzis, George. *In Letters of Blood and Fire: Work, Machines, and the Crisis of Capitalism*. Oakland, CA: PM Press, 2012.

Campbell-Kelly, Martin. *From Airline Reservations to Sonic the Hedgehog: A History of the Software Industry*. Cambridge, MA: MIT Press, 2003.

Chen, Adrian. "The Laborers Who Keep the Dick Pics and Beheadings out of Your Facebook Feed." *Wired*, October 23, 2014. https://www.wired.com/2014/10/content-moderation/.

Chun, Wendy H. K. *Programmed Visions: Software and Memory*. Cambridge, MA: MIT Press, 2011.

Cowen, Deborah. *The Deadly Life of Logistics: Mapping Violence in Global Trade*. Minneapolis: University of Minnesota Press, 2014.

Crogan, Patrick. *Gameplay Mode: War, Simulation, and Technoculture*. Minneapolis: University of Minnesota Press, 2011.

Cuppini, Niccolò, Mattia Frapporti, and Maurilio Pirone. "Logistics Struggles in the Po Valley Region: Territorial Transformations and Processes of Antagonistic Subjectivation." *South Atlantic Quarterly* 114, no. 1 (2015): 119–34.

Dachwitz, Ingo, and Markus Reuter. "Warum Künstliche Intelligenz Facebooks Moderationsprobleme nicht lösen kann, ohne neue zu schaffen." *Netzpolitik*, May 4, 2019. https://netzpolitik.org/2019/warum-kuenstliche-intelligenz-facebooks-moderationsprobleme-nicht-loesen-kann-ohne-neue-zu-schaffen/.

Dibbell, Julian. "The Chinese Game Room: Play, Productivity, and Computing at Their Limits." *Artifact* 2, no. 2 (2008): 82–87.

Dibbell, Julian. "The Decline and Fall of an Ultra Rich Online Gaming Empire." *Wired*, November 24, 2008. https://www.wired.com/2008/11/ff-ige/.

Dodge, Martin, and Rob Kitchin. "Codes of Life: Identification Codes and the Machine-Readable World." *Environment and Planning D: Society and Space* 23, no. 6 (2005): 851–81.

Dodge, Martin, and Rob Kitchin. *Code/Space: Software and Everyday Life*. Cambridge, MA: MIT Press, 2011.

Dwoskin, Elizabeth, Jeanne Whalen, and Regine Cabato. "Content Moderators at YouTube, Facebook and Twitter See the Worst of the Web." *Washington Post*, July 25, 2019. https://www.washingtonpost.com/technology/2019/07/25/social-media-companies-are-outsourcing-their-dirty-work-philippines-generation-workers-is-paying-price/.

Dyer-Witheford, Nick. *Cyber-Marx: Cycles and Circuits of Struggle in High-Technology Capitalism*. Urbana: University of Illinois Press, 1999.

Dyer-Witheford, Nick. *Cyber-Proletariat: Global Labour in the Digital Vortex*. London: Pluto Press, 2015.

Dyer-Witheford, Nick. "Empire, Immaterial Labor, the New Combinations, and the Global Worker." *Rethinking Marxism* 13, no. 3–4 (2001): 70–80.

Dyer-Witheford, Nick, and Greig de Peuter. *Games of Empire: Global Capitalism and Video Games*. Minneapolis: University of Minnesota Press, 2009.

Dyer-Witheford, Nick, and Greig de Peuter. "Postscript: Gaming While Empire Burns." *Games and Culture* 16, no. 3 (2020): 371–80.

Easterling, Keller. *Extrastatecraft: The Power of Infrastructure Space*. London: Verso, 2014.

Economist, The. 2015. "Digital Taylorism. A Modern Version of 'Scientific Management' Threatens to Dehumanise the Workplace." *The Economist*, September 10, 2015. https://www.economist.com/business/2015/09/10/digital-taylorism.

Elegant, Naomi Xu. "The Internet Cloud Has a Dirty Secret." *Fortune*, September 18, 2019. https://fortune.com/2019/09/18/internet-cloud-server-data-center-energy-consumption-renewable-coal/.

Fang, Lee. "Google Hired Gig Economy Workers to Improve Artificial Intelligence in Controversial Drone Targeting Project." *The Intercept*, February 4, 2019. https://theintercept.com/2019/02/04/google-ai-project-maven-figure-eight/.

Frank, Michael. "How Telematics Has Completely Revolutionized the Management of Fleet Vehicles." *Entrepreneur Europe*, October 20, 2014. https://www.entrepreneur.com/article/237453.

Freeman, Joshua B. *Behemoth: A History of the Factory and the Making of the Modern World*. New York: W. W. Norton, 2018.

Fuchs, Christian. *Digital Labor and Karl Marx*. New York: Routledge, 2014.

Fuller, Matthew, ed. *Software Studies: A Lexicon*. Cambridge, MA: MIT Press, 2008.

Fuller, Matthew, and Andrew Goffey. *Evil Media*. Cambridge, MA: MIT Press, 2012.

Gabrys, Jennifer. *Digital Rubbish: A Natural History of Electronics*. Ann Arbor: University of Michigan Press, 2013.

Gerlitz, Carolin, and Anne Helmond. "The Like Economy: Social Buttons and the Data-Intensive Web." *New Media & Society* 15, no. 8 (2013): 1348–65.

Gilbert, David. "Facebook Is Forcing Its Moderators to Log Every Second of Their Days—Even in the Bathroom." *Vice*, January 9, 2020. https://www.vice.com/en/article/z3beea/facebook-moderators-lawsuit-ptsd-trauma-tracking-bathroom-breaks.

Glanz, James. "Power, Pollution and the Internet." *New York Times*, September 22, 2012. http://www.nytimes.com/2012/09/23/technology/data-centers-waste-vast-amounts-of-energy-belying-industry-image.html.

Gorißen, Stefan. "Fabrik." In *Enzoklpädie der Neuzeit, Bd. 3*, edited by Friedrich Jaeger, 740–47. Stuttgart: J. B. Metzler, 2006.

Graham, Mark, and Mohammad Anwar. "The Global Gig Economy: Towards a Planetary Labour Market?" *First Monday* 24, no. 4 (2019).

Graham, Mark, Isis Hjorth, and Vili Lehdonvirta. "Digital Labour and Development: Impacts of Global Digital Labour Platforms and the Gig Economy on Worker Livelihoods." *Transfer: European Review of Labour and Research* 23, no. 2 (2017): 135–62.

Graham, Mark, and Laura Mann. "Imagining a Silicon Savannah? Technological and Conceptual Connectivity in Kenya's BPO and Software Development Sectors." *Electronic Journal of Information Systems in Developing Countries* 56, no. 1 (2013): 1–19.

Grassegger, Hannes, and Till Krause. "Im Netz des Bösen." *Süddeutsche Zeitung Magazin*, December 15, 2016. http://www.sueddeutsche.de/digital/inside-facebook-im-netz-des-boesen-1.3295206.

Gray, Mary L., and Siddharth Suri. *Ghost Work: How to Stop Silicon Valley from Building a New Global Underclass*. Boston: Houghton Mifflin Harcourt, 2019.

Greenpeace. *Make IT Green. Cloud Computing and its Contribution to Climate Change*. Amsterdam: Greenpeace International, 2010.

Grier, David A. *When Computers Were Human*. Princeton, NJ: Princeton University Press, 2013.

Guth, Robert A., and Nick Wingfield. "Workers at EA Claim They Are Owed Overtime." *Wall Street Journal*, November 19, 2004. https://www.wsj.com/articles/SB110081756477478548.

Hardt, Michael, and Antonio Negri. *Empire*. Cambridge, MA: Harvard University Press, 2000.

Harney, Stefano, and Fred Moten. *The Undercommons: Fugitive Planning and Black Study*. New York: Minor Compositions, 2013.

Harvey, David. *Spaces of Capital: Towards a Critical Geography*. New York: Routledge, 2001.

Head, Simon. *Mindless: Why Smarter Machines Are Making Dumber Humans*. New York: Basic Books, 2014.

Head, Simon. *The New Ruthless Economy: Work and Power in the Digital Age*. New York: Oxford University Press, 2005.

Heeks, Richard. "Current Analysis and Future Research Agenda on 'Gold Farming': Real-World Production in Developing Countries for the Virtual Economies of Online Games." Development Informatics Working Paper. Manchester: Institute for Development Policy and Management, 2008.

Heeks, Richard. "Decent Work and the Digital Gig Economy: A Developing Country Perspective on Employment Impacts and Standards in Online Outsourcing, Crowdwork, Etc." Development Informatics Working Paper. Manchester: Institute for Development Policy and Management, 2017.

Heeks, Richard. "How Many Platform Workers Are There in the Global South?" *ICTs for Development*, January 29, 2019. https://ict4dblog.wordpress.com/2019/01/29/how-many-platform-workers-are-there-in-the-global-south/.

Heeks, Richard. "Understanding 'Gold Farming' and Real-Money Trading as the Intersection of Real and Virtual Economies." *Journal for Virtual Worlds Research* 2, no. 4 (2009): 1–27.

Helmond, Anne, David B. Nieborg, and Fernando N. van der Vlist. "Facebook's Evolution: Development of a Platform-as-Infrastructure." *Internet Histories* 3, no. 2 (2019): 123–46.

Holmes, Brian. "Do Containers Dream of Electric People? The Social Form of Just-in-Time Production." *Open*, no. 21 (2011): 30–44.

Hu, Tung-Hui. *A Prehistory of the Cloud*. Cambridge, MA: MIT Press, 2015.

Huifeng, He. "Chinese 'Farmers' Strike Cyber Gold." *South China Morning Post*, October 25, 2005. https://www.scmp.com/node/521571.

Hürtgen, Stefanie, Boy Lüthje, Wilhelm Schumm, and Martina Sproll. *Von Silicon Valley nach Shenzhen: Globale Produktion und Arbeit in der IT-Industrie*. Hamburg: VSA, 2009.

Huws, Ursula. *Labor in the Global Digital Economy: The Cybertariat Comes of Age*. New York: Monthly Review Press, 2014.

Huws, Ursula. "Logged Labour: A New Paradigm of Work Organisation?" *Work Organisation, Labour and Globalisation* 10, no. 1 (2016): 7–26.

Huws, Ursula. *The Making of a Cybertariat: Virtual Work in a Real World*. New York: Monthly Review Press, 2003.

Huws, Ursula, Neil Spencer, Matthew Coates, Dag Sverre Syrdal, and Kaire Holts. *The Platformisation Of Work in Europe: Results from Research in 13 European Countries*. Brussels: Foundation for European Progressive Studies (FEPS), 2019.

Irani, Lilly. "Difference and Dependence among Digital Workers: The Case of Amazon Mechanical Turk." *South Atlantic Quarterly* 114, no. 1 (2015): 225–34.

Irani, Lilly. "Justice for 'Data Janitors.'" *Public Books*, January 15, 2015. http://www.publicbooks.org/justice-for-data-janitors/.

Jacobs, Harrison. "Inside 'iPhone City,' the Massive Chinese Factory Town Where Half of the World's iPhones Are Produced." *Business Insider*, May 7, 2018. https://www.businessinsider.com/apple-iphone-factory-foxconn-china-photos-tour-2018-5.

Jones, Nicola. "How to Stop Data Centres from Gobbling up the World's Electricity." *Nature*, September 12, 2018. https://www.nature.com/articles/d41586-018-06610-y.

Kaplan, Esther. "The Spy Who Fired Me." *Harper's*, March 2015. http://harpers.org/archive/2015/03/the-spy-who-fired-me/2/.

Kitchin, Rob. "Thinking Critically about and Researching Algorithms." *Information, Communication & Society* 20, no. 1 (2017): 14–29.

Kline, Stephen, Nick Dyer-Witheford, and Greig De Peuter. *Digital Play: The Interaction of Technology, Culture and Marketing*. Montreal: McGill-Queen's University Press, 2003.

Koebler, Jason, and Joseph Cox. "The Impossible Job: Inside Facebook's Struggle to Moderate Two Billion People." *Vice*, August 23, 2018. https://www.vice.com/en_us/article/xwk9zd/how-facebook-content-moderation-works.

Kucklick, Christoph. "SMS-Adler." *Brandeins*, no. 4 (2011): 26–34.

Kuek, Siou Chew, Cecilia Paradi-Guilford, Toks Fayomi, Saori Imaizumi, Panos Ipeirotis, Patricia Pina, and Manpreet Singh. *The Global Opportunity in Online Outsourcing*. Washington, DC: World Bank, 2015.

Lazzarato, Maurizio. "Immaterial Labour." In *Radical Thought in Italy: A Potential Politics*, edited by Paulo Virno and Michael Hardt, 133–47. Minneapolis: University of Minnesota Press, 1996.

LeCavalier, Jesse. *The Rule of Logistics: Walmart and the Architecture of Fulfillment*. Minneapolis: University of Minnesota Press, 2016.

Lecher, Colin. "How Amazon Automatically Tracks and Fires Warehouse Workers for 'Productivity.'" *The Verge*, April 25, 2019. https://www.theverge.com/2019/4/25/18516004/amazon-warehouse-fulfillment-centers-productivity-firing-terminations.

Lichtenstein, Nelson. *The Retail Revolution: How Walmart Created a Brave New World of Business*. New York: Picador, 2009.

Light, Jennifer S. "When Computers Were Women." *Technology and Culture* 40, no. 3 (1999): 455–83.

Loomis, Carol J. "The Sinking of Bethlehem Steel." *Fortune*, April 5, 2004. http://archive.fortune.com/magazines/fortune/fortune_archive/2004/04/05/366339/index.htm.

Lund, John, and Christopher Wright. "State Regulation and the New Taylorism: The Case of Australian Grocery Warehousing." *Relations Industrielles/Industrial Relations* 56, no. 4 (2001): 747–69.

Lüthje, Boy. *Standort Silicon Valley: Ökonomie und Politik der vernetzten Massenproduktion*. Frankfurt am Main: Campus, 2001.

Lyster, Claire. *Learning from Logistics: How Networks Change our Cities*. Basel: Birkhäuser, 2016.

Marx, Karl. *Capital: A Critique of Political Economy*. Vol 1. London: Penguin, 2004.

Marx, Karl. *Capital: A Critique of Political Economy*. Vol 2. London: Penguin, 1992.

Marx, Karl. *Grundrisse: Foundations of the Critique of Political Economy*. London: Penguin, 2005.

Matuschek, Ingo, Kathrin Arnold, and Günther G. Voß. *Subjektivierte Taylorisierung*. Munich: Rainer Hampp, 2007.

Mezzadra, Sandro. "The Gaze of Autonomy: Capitalism, Migration, and Social Struggles." In *The Contested Politics of Mobility: Borderzones and Irregularity*, edited by Vicki Squire, 121–42. London: Routledge, 2011.

Mezzadra, Sandro, and Brett Neilson. *Border as Method, or, the Multiplication of Labor*. Durham, NC: Duke University Press, 2013.

Mezzadra, Sandro, and Brett Neilson. *The Politics of Operations: Excavating Contemporary Capitalism*. Durham, NC: Duke University Press, 2019.

Mitchell, Mariah. "I Deliver Your Food. Don't I Deserve Basic Protections?" *New York Times*, March 17, 2020. https://www.nytimes.com/2020/03/17/opinion/coronavirus-food-delivery-workers.html.

Moore, Jason W., ed. *Anthropocene or Capitalocene?: Nature, History, and the Crisis of Capitalism*. Oakland, CA: PM Press, 2016.

Moore, Jason W. *Capitalism in the Web of Life: Ecology and the Accumulation of Capital*. London: Verso, 2015.

Mosco, Vincent. *To the Cloud: Big Data in a Turbulent World*. Boulder, CO: Paradigm, 2015.

Nachtwey, Oliver, and Philipp Staab. "Die Avantgarde des Digitalen Kapitalismus." *Mittelweg* 36, no. 6 (December 2015–January 2016): 59–84.

Nakamura, Lisa. "Don't Hate the Player, Hate the Game: The Racialization of Labor in World of Warcraft." *Critical Studies in Media Communication* 26, no. 2 (2009): 128–44.

Nechushtai, Efrat. "Could Digital Platforms Capture the Media through Infrastructure?" *Journalism* 19, no. 8 (2018): 1043–58.

Negri, Antonio. *Goodbye Mr. Socialism. In Conversation with Raf Valvola Scelsi*. New York: Seven Stories Press, 2008.

Neilson, Brett, and Tanya Notley. "Data Centres as Logistical Facilities: Singapore and the Emergence of Production Topologies." *Work Organisation, Labour & Globalisation* 13, no. 1 (2019): 15–29.

Nelson, Benjamin. *Punched Cards to Bar Codes: A 200 Year Journey*. Chicago: Helmers, 1997.

Newton, Casey. "The Terror Queue." *The Verge*, December 16, 2019. https://www.theverge.com/2019/12/16/21021005/google-youtube-moderators-ptsd-accenture-violent-disturbing-content-interviews-video.

Ngai, Pun. *Migrant Labor in China*. Cambridge: Polity, 2016.

Ngai, Pun, and Ralf Ruckus. *iSlaves: Ausbeutung und Widerstand in Chinas Foxconn-Fabriken*. Vienna: Mandelbaum, 2013.

Nieborg, David B, and Anne Helmond. "The Political Economy of Facebook's Platformization in the Mobile Ecosystem: Facebook Messenger as a Platform Instance." *Media, Culture & Society* 41, no. 2 (2019): 196–218.

Noble, David. *Digital Diploma Mills: The Automation of Higher Education*. New York: Monthly Review Press, 2003.

O'Connor, Sarah. "Amazon Unpacked." *Financial Times*, February 8, 2013. https://www.ft.com/content/ed6a985c-70bd-11e2-85d0-00144feab49a.

Parikka, Jussi. "Cultural Techniques of Cognitive Capitalism: Metaprogramming and the Labour of Code." *Cultural Studies Review* 20, no. 1 (2014): 30–52.

Parisi, Luciana. *Contagious Architecture: Computation, Aesthetics, and Space*. Cambridge, MA: MIT Press, 2013.

Parks, Lisa, and James Schwoch, eds. *Down to Earth: Satellite Technologies, Industries, and Cultures*. New Brunswick, NJ: Rutgers University Press, 2012.

Parks, Lisa, and Nicole Starosielski. *Signal Traffic: Critical Studies of Media Infrastructures*. Urbana: University of Illinois Press, 2015.

Pasquinelli, Matteo. "Google's PageRank Algorithm: A Diagram of the Cognitive Capitalism and the Rentier of the Common Intellect." In *Deep Search*, edited by Felix Stalder and Konrad Becker, 152–62. Innsbruck: Studienverlag, 2009.

Pellow, David N., and Lisa Sun-Hee Park. *The Silicon Valley of Dreams: Environmental Injustice, Immigrant Workers, and the High-Tech Global Economy*. New York: New York University Press, 2002.

Pias, Claus. "Computer Spiel Welten." PhD diss., Bauhaus-Universität Weimar, 2000.

Punsmann, Burcu Gültekin. "Three Months in Hell." *Süddeutsche Zeitung Magazin*, January 6, 2018. https://sz-magazin.sueddeutsche.de/internet/three-months-in-hell-84381.

Puzder, Andy. "The Minimum Wage Should Be Called the Robot Employment Act." *Wall Street Journal*, April 3, 2017. https://www.wsj.com/articles/the-minimum-wage-should-be-called-the-robot-employment-act-1491261644.

Raffetseder, Eva-Maria, Simon Schaupp, and Philipp Staab. "Kybernetik und Kontrolle. Algorithmische Arbeitssteuerung und Betriebliche Herrschaft." *PROKLA. Zeitschrift für kritische Sozialwissenschaft* 47, no. 2 (2017): 229–48.

Reifer, Thomas. "Unlocking the Black Box of Globalization." Paper presented at The Travelling Box: Containers as the Global Icon of our Era, University of California, Santa Barbara, November 8–10, 2007.

Rettberg, Scott. "Corporate Ideology in World of Warcraft." In *Digital Culture, Play, and Identity. A World of Warcraft Reader*, edited by Hilde G. Corneliussen and Jill W. Rettberg, 19–38. Cambridge, MA: MIT Press, 2008.

Reuter, Markus, Ingo Dachwitz, Alexander Fanta, and Markus Beckedahl. "Exklusiver Einblick: So funktionieren Facebooks Moderationszentren." *Netzpolitik*, April 5, 2019. https://netzpolitik.org/2019/exklusiver-einblick-so-funktionieren-facebooks-moderationszentren/.

Roberts, Sarah T. *Behind the Screen: Content Moderation in the Shadows of Social Media*. New Haven, CT: Yale University Press, 2019.

Roberts, Sarah T. "Digital Refuse: Canadian Garbage, Commercial Content Moderation and the Global Circulation of Social Media's Waste." *Media Studies Publications* 10 (2016). https://ir.lib.uwo.ca/cgi/viewcontent.cgi?article=1014&context=commpub/.

Rossiter, Ned. *Software, Infrastructure, Labor: A Media Theory of Logistical Nightmares*. New York: Routledge, 2016.

Rudacille, Deborah. "In Baltimore, Visions of Life after Steel." *CityLab*, May 15, 2019. https://www.citylab.com/life/2019/05/bethlehem-steel-mill-photos-sparrows-point-dundalk-baltimore/589465/.

Ruffino, Paolo, and Jamie Woodcock. "Game Workers and the Empire: Unionisation in the UK Video Game Industry." *Games and Culture* 16, no. 3 (2020): 317–28.

Sadowski, Jathan. "Cyberspace and Cityscapes: On the Emergence of Platform Urbanism." *Urban Geography* 41, no. 3 (2020): 448–52.

Satariano, Adam. "How the Internet Travels across Oceans." *New York Times*, March 10, 2019. https://www.nytimes.com/interactive/2019/03/10/technology/internet-cables-oceans.html.

Schmidt, Florian A. *Crowdproduktion von Trainingsdaten: Zur Rolle von Online-Arbeit beim Trainieren autonomer Fahrzeuge*. Düsseldorf: Hans-Böckler-Stiftung, 2019.

Scholz, Trebor, ed. *Digital Labor: The Internet as Playground and Factory*. New York: Routledge, 2013.

Scholz, Trebor, and Nathan Schneider, eds. *Ours to Hack and to Own: The Rise of Platform Cooperativism, a New Vision for the Future of Work and a Fairer Internet*. New York: OR Books, 2016.

Sekula, Allan. *Fish Story*. Düsseldorf: Richter, 2002.

Shane, Scott. "Prime Mover: How Amazon Wove Itself into the Life of an American City." *New York Times*, November 30, 2019. https://www.nytimes.com/2019/11/30/business/amazon-baltimore.html.

Silberman, M. Six, and Lilly Irani. "Operating an Employer Reputation System: Lessons from Turkopticon, 2008–2015." *Comparative Labor Law & Policy Journal* 37, no. 3 (2016): 505–41.

Smith, Rebecca, Paul Alexander Marvy, and Jon Zerolnick. *The Big Rig Overhaul. Restoring Middle-Class Jobs at America's Ports through Labor Law Enforcement*. New York: National Employment Law Project, 2014.

Staab, Philipp. *Digitaler Kapitalismus: Markt und Herrschaft in der Ökonomie der Unknappheit*. Berlin: Suhrkamp, 2019.

Starosielski, Nicole. *The Undersea Network*. Durham, NC: Duke University Press, 2015.

Statt, Nick, Casey Newton, and Zoe Schiffer. "Facebook Moderators at Accenture Are Being Forced Back to the Office, and Many Are Scared for Their Safety." *The Verge*, October 1, 2020. https://www.theverge.com/2020/10/1/21497789/facebook-content-moderators-accenture-return-office-coronavirus.

Taylor, Frederick W. *The Principles of Scientific Management*. New York: Cosimo, 2010. Originally published in 1911 by Harper and Brothers.

Terranova, Tiziana. "Free Labor: Producing Culture for the Digital Economy." *Social Text* 18, no. 2 (2000): 33–58.

Terranova, Tiziana. "Red Stack Attack! Algorithms, Capital, and the Automation of the Common." In *#Accelerate: The Accelerationist Reader*, edited by Robin Mackay and Armen Avanessian, 379–97. Falmouth, UK: Urbanomic, 2014.

Thompson, Paul, Rachel Parker, and Stephen Cox. "Interrogating Creative Theory and Creative Work: Inside the Games Studio." *Sociology* 50, no. 2 (2016): 316–32.

Thrift, Nigel. *Knowing Capitalism*. London: Sage, 2005.

Toscano, Alberto. "Factory, Territory, Metropolis, Empire." *Angelaki* 9, no. 2 (2004): 197–216.

Toscano, Alberto, and Jeff Kinkle. *Cartographies of the Absolute*. London: Zed Books.

Tronti, Mario. *Arbeiter und Kapital*. Frankfurt: Neue Kritik, 1974.

Tsing, Anna. "Supply Chains and the Human Condition." *Rethinking Marxism* 21, no. 2, (2009): 148–76.

Uhl, Karsten. "Work Spaces: From the Early-Modern Workshop to the Modern Factory Workshop and Factory." *European History Online (EGO)*. February 5, 2016. http://www.ieg-ego.eu/uhlk-2015-en.

Virno, Paolo. "General Intellect." *Historical Materialism* 15, no. 3 (2007): 3–8.

Virno, Paolo. *A Grammar of the Multitude: For an Analysis of Contemporary Forms of Life*. Los Angeles: Semiotext(e), 2004.

Wallace, Rob. *Dead Epidemiologists: On the Origins of Covid-19*. New York: Monthly Review Press, 2020.

Williams, Ian. "'You Can Sleep Here All Night': Video Games and Labor." *Jacobin*, November 8, 2013. https://jacobinmag.com/2013/11/video-game-industry/.

Woodcock, Jamie. "The Work of Play: Marx and the Video Games Industry in the United Kingdom." *Journal of Gaming & Virtual Worlds* 8, no. 2 (2016): 131–43.

Woodcock, Jamie, and Mark Graham. *The Gig Economy: A Critical Introduction*. Cambridge: Polity, 2019.

INDEX

Ingram Content Group UK Ltd.
Milton Keynes UK
UKHW021948300523
422595UK00005B/233